Fourth Edition

BEHAVIORAL SCIENCE
Statistics

PJ Verrecchia

Kendall Hunt
publishing company

Cover image © Shutterstock.com

Kendall Hunt
publishing company

www.kendallhunt.com
Send all inquiries to:
4050 Westmark Drive
Dubuque, IA 52004-1840

Published in the United States of America

TABLE OF CONTENTS

Behavioral Science Statistics is a textbook appropriate for introductory-level statistics courses. It is written in an easy to understand style, yet it thoroughly explains statistical concepts. The book contains numerous practical examples as well as step-by-step instructions on how to perform each statistical analysis. There are practice problems at the end of each chapter for readers to hone their newly acquired skills. The textbook author, Dr. P. J. Verrecchia, has been teaching statistics since 2000 for numerous college majors, including Criminal Justice, Behavioral Sciences, Nursing, and Psychology. **Behavioral Science Statistics** is a textbook that crosses disciplines and is useful in any number of statistics classes.

ACKNOWLEDGMENTS

First and foremost I would like to thank God for giving me the ability and perseverance to write this book. "For from Him and through Him are all things."—Romans 11:36.

A very supportive family has always been a blessing for me. My mother, Jeannine Naftzger, has always been a role model without peer. A loving brother, Andrew, and sisters, Nini McGlinchey and Dale Carslaw, and my nieces and nephews keep me motivated and grounded. My aunt and uncle, Helen and Tom Wallace, spent many years of their lives as excellent teachers and taught me the value of hard work and education. My brother in law (I could not have asked for a better one), Sergeant Anthony Genaro of the Cape May, New Jersey, Police Department, also deserves my thanks for his help with this book.

Over my many years of teaching I have been blessed to have met many outstanding students, but one deserves special mention. Kirsten Hutzell, PhD, and I met in 2006 when she was an undergraduate student of mine at York College of Pennsylvania. She is now an Assistant Professor of Criminology and Criminal Justice at York College of Pennsylvania, and I could not be more proud. She deserves special mention for her contributions to this book.

God has also blessed me with outstanding teachers. John H. Lemmon, PhD, has always been not just my teacher but my intellectual mentor, and I am forever grateful for his friendship. Thanks also to George T. Haskett, PhD, and Jeffrey D. Leitzel, PhD, for being outstanding teachers and mentors.

Nicole Mathers, Brenda Rolwes, Rene Caroline Balan, Tessa Ruchotzke, Torrie Johnson, and Nick Carolan from Kendall Hunt Publishing also deserve my thanks and praise for their help and patience with this project.

Finally, this book would never have come to fruition without the help and patience of my wife, Jennifer, and daughters Sarah and Sophie.

DEDICATION

*This book is dedicated to my wife
Jennifer and daughters Sarah and Sophie.
You are my sources of constant joy and
inspiration. Statistically speaking,
you are the greatest.*

CHAPTER 1

INTRODUCTION TO STATISTICS

© Andrii Kondiuk/Shutterstock.com

You are finally being forced to take a statistics course. Of course, that sentence requires some assumptions on my part. One, that this is a required course. Two, that you have been putting this off until you absolutely had to take a statistics class. And three, that you are not looking forward to this experience.

Over the years that I have taught statistics I have noticed what I call "statistics apprehension," and most of that apprehension comes from a dread of math. While there is a great deal of math in statistics, this is more than a math course. You will have to calculate answers using formulas, but this book will emphasize knowing how to interpret that calculation. You see, you may never need to calculate another statistic after you use this book (unless you are going to graduate school, in which case you will calculate many). But regardless of the field you enter, you will have to know how to think statistically.

We live in a data-driven world. Data (the plural of the word datum) are the scores obtained in research based on observations made in our environment and can be represented by numbers. Just pick up a newspaper (or read a news report online) and you will see many statistics: the number of people who engaged in a recent protest; the increase (or decrease) of income in an area; the chance of rain. One reason this information is represented in numbers is that numbers are more precise and less ambiguous than words. For example, it is more exact for a newspaper report to say that there were 150 people at a protest than it is to say that there were a lot of people at a protest.

The person who can think in terms of numbers is the person who will dramatically increase their chances of success, regardless of their field of endeavor. Supreme Court Justice Oliver Wendell Holmes in 1897 said, "The man of the future is the man of statistics." Statistics forms an incredibly important part of the Behavioral Sciences. Whenever you hear terms like "data-driven approach" or "best practices," you are hearing references to making decisions based on hard data. In your professional life, you will see (and perhaps even fill out) performance evaluations, tables, graphs, and charts. This book will help you make sense of them and all of the numbers you will encounter in life. When you read a research-based report (and you will), you will need to understand how the person (or people) who wrote that report came to their conclusions so you can make an informed assessment about those conclusions.

But, oh, the math!

We will get to that, but first I want to go over some basic terms that are used in statistics and research, because we will be using them throughout this book.

BASIC TERMS AND CONCEPTS

Research

Research is a systematic way to observe and understand the social world. It is more than casual observation—research makes us pay attention. To say that something is understood using research means that we are following a set of steps which include selecting a research topic, reviewing the relevant literature, selecting a method of observation, making and recording those observations, analyzing the data, and communicating the results. Statistics provides us with a quantitative method for making and recording those observations.

© Rawpixel.com/Shutterstock.com

Variables

In research, we speak in variables. That was not a typo. We do not speak *of* variables but *in* variables. Variables are concepts that, when measured, have two or more values. For example, attitude towards the death penalty is a variable because some people are in favor of the death penalty, some are against

© Casimiro PT/Shutterstock.com

the death penalty, and some do not care either way. In other words, attitude toward the death penalty varies from person to person.

One type of variable is the **independent variable**, which is also referred to as the causal variable. The independent variable causes another variable, which is the **dependent variable**—the effect variable. For example, the number of prior convictions someone has (independent variable) has an effect on the length of their sentence (dependent variable) if they get another conviction. As you can see, we are using numbers to determine a relationship (number of prior offenses and length of sentence). It is assumed that if someone has more priors, they will receive a longer sentence.

This is assumed because of a third variable: the **intervening variable**. This is a variable that comes between the independent variable and dependent variable and has an effect on the relationship. For example, the variable type of attorney (private attorney, public defender, or self-representation) someone has can have an effect on the length of the sentence they get. Of course, that is affected by another intervening variable—socio-economic status—which demonstrates how complex relationships can be. We use statistics to help make sense of these relationships, and the social world in general.

Attributes

When I said that variables have two or more values when measured, we refer to those values as attributes. Hence, variables are logical groupings of attributes. For example, the attributes for year in college are freshman, sophomore (fun fact, sophomore means wise fool), junior, senior, and graduate student. The attributes for a variable should be exhaustive, which means that all attributes for that variable should be represented.

So, we can measure attitude toward the death penalty a couple of ways. One would be to ask people "Are you against the death penalty?" and have the following categories (attributes) for their response: Yes, No, or No opinion (leaving out No opinion would not make our attributes exhaustive because, as I said earlier, some people do not care one way or the other). Another way to measure attitude toward the death penalty would be to use a Likert scale, which would follow a statement like: "I am in favor of the death penalty" with the attributes Strongly Disagree, Disagree, No Opinion, Agree, and Strongly Agree. In each instance, we are measuring the variable attitude toward the death penalty, but we are using different attributes.

Theory

Behavioral Sciences research focuses on what is, as opposed to what should be. **Theory** plays a very important role in Behavioral Sciences research, whether we are testing a theory (known as applied research) or developing a theory (known as basic research). Theory provides an explanatory framework for our observations. For example, a part of Clifford Shaw and Henry D. McKay's

THEORY
COMMUNICATIONS

Social Disorganization Theory (1942) says that population instability (people moving in and out of an area) contributes to juvenile delinquency. In their theory, Shaw and McKay were trying to explain delinquency (in part) due to the number of people moving in and out of an area and not putting down roots.

Hypothesis

A **hypothesis** is not an educated guess. It is a tentative statement about a relationship between variables. The reason it is a tentative statement is that it has yet to be tested. So, a hypothesis for the population instability aspect of the Social Disorganization Theory would be that an area with greater population instability (independent variable) will have a higher rate of delinquency (dependent variable) than an area with less population instability.

© Dizain/Shutterstock.com

TYPES OF STATISTICS

In this book we will learn about two types of statistics, **descriptive** and **inferential**. To understand the difference between them is to understand the difference between samples and populations. A **population** is a group that you want to learn more about, for example, college students in the United States. Since there are millions of college students in the United States, it would be impossible to conduct research with all of them. So we take a **sample** of the population, which is a subset of the population, and that is the group we actually study. If you use correct sampling techniques, your sample should be representative of the population. (We will talk more about representativeness in the chapter on probability.) These two types of statistics relate to the four purposes of research.

© Phipatbig/Shutterstock.com

PURPOSES OF RESEARCH

Years ago, someone came to my office and said that he needed my help. He told me that there was a crime problem in his town and he wanted me to recommend to his town council that they should have a citizen's bike patrol program, and I told him that I would look into his town's crime problem and get back to him before I made any recommendations.

Before I could make any recommendations, I needed to research whether his town, in fact, had a crime problem. Another professor joined me, and we recruited two students to be a part of the team, and we went and collected data. We looked at Federal Bureau of Investigation Uniform Crime Report (UCR) data trends for the town over the previous 10 years. We surveyed residents in the town, and we conducted focus group interviews with business owners in the town.

Exploration is the first purpose of research. In the exploratory phase of our project, we were using descriptive data—data that described the current crime problem in the town. The second purpose of research is **description**. After we collected our data, we compiled a report of our findings. We used descriptive statistics to demonstrate what our findings were. Our report included graphs, tables, and summaries of our findings. In the descriptive phase of research, the researcher is using descriptive statistics to make a visual representation (paint a picture, if you will) of their findings. What we found was that the town in question did not have a crime problem (according to residents, lack of parking was a bigger problem than crime), so we said that we could not support using taxpayer money to fund a citizen's bike patrol program.

The third purpose of research is **explanation**, which is trying to determine why something happens. Descriptive statistics can demonstrate that there is a problem (such as high crime in a certain area), but we use inferential statistics to explain why. This would happen if a certain pattern could be discerned in the sample and then applied to the population. The final purpose of research is **application**, which is where we use research to answer a question or solve a problem. This is done sometimes with a new program, say, truancy reduction. Applied research would evaluate the program and look at (among other things), truancy rates before initiation of the program and after.

THE MATH USED IN STATISTICS

Many students dread taking statistics because they are math phobic. There is no need for this, because the math used in statistics involves addition ($+$), subtraction ($-$), multiplication ($\times$), division ($\div$), squaring a number (multiplying it by itself, or n^2), and taking the square root of a number ($\sqrt{n}$). That is all there is to it. If you have a calculator that can perform these functions (and you will need a calculator to take statistics), you will master the math in statistics.

To demonstrate, let us look at a formula we will use later in the book, the Pearson's Product Moment Correlation Coefficient (commonly known as Pearson's r):

$$r = \frac{N\left(\sum XY\right) - \left(\sum X\right)\left(\sum Y\right)}{\sqrt{\left[N\left(\sum X^2\right) - \left(\sum X\right)^2\right]\left[N\left(\sum Y^2\right) - \left(\sum Y\right)^2\right]}}$$

While this formula may look daunting, let us look at the math involved. You will use all of the mathematical operations I listed in the above paragraph to solve this equation. The Greek Σ means sum, which means to add. Obviously the subtraction signs mean to subtract, and putting numbers next to each other (or parentheses next to each other) means we multiply them. The long line separating the top half of the equation from the bottom half means to divide, the superscript ^{2} means to square (multiply the number by itself), and the square root signs indicates that we take the square root of the denominator. So, as complex as this formula looks, it is solved through addition, subtraction, multiplication, division, squaring, and taking a square root.

Order of Operations

When we were young, we learned the order of mathematical operations as PEMDAS, or by the mnemonic Please Excuse My Dear Aunt Sally. This translates as Parentheses, Exponents, Multiplication, Division, Addition, and Subtraction.

Compute all numbers in parentheses first. Brackets may replace parentheses, but you may see parentheses inside of brackets. If there are parentheses inside of brackets, compute the innermost parentheses first:

$$8 \times [5 + (2 \times 6)]$$
$$= 8 \times (5 + 12)$$
$$= 8 \times 17$$
$$= 136$$

Exponents are next, and they are numbers raised above and to the right of another number. They tell us how many times to multiply that number by itself. In statistics, the only exponent we use is 2, which means we will only be squaring numbers. So 5^2 means multiply 5 times itself so $5^2 = 5 \times 5 = 25$.

$$6 + (8 \times 4) + (7 - 3^2)$$
$$= 6 + 32 + (7 - 9)$$
$$= 6 + 32 + (-2)$$
$$= 38 + (-2)$$
$$= 36$$

After parentheses and exponents have been computed, multiplication and division (from left to right) are next. Addition and subtraction (also from left to right) are performed last.

$$(6 + 5^2) \times 3 + [(4 \times 7) \div 14]$$
$$= (6 + 25) \times 3 + (28 \div 14)$$
$$= 31 \times 3 + 2$$
$$= 93 + 2$$
$$= 95$$

Solving Formulas

In statistics we use formulas to find an answer, and in formulas we use symbols to stand for that answer. For example, in the formula $Y' = b(x) + a$, the Y' stands for the answer that we obtain. (We will see this formula again in the chapter on Logistic Regression.) We do not know the answer (value for Y'), but we do know the values of b, x and a. The first thing we do is rewrite the formula, replacing the letters with their numerical value. So in this example if $b = 5$, $x = 2$, and $a = 11$, rewriting the formula gives us

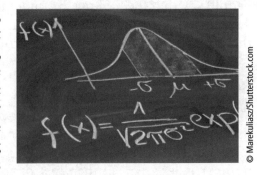

© Marekuliasz/Shutterstock.com

$$Y' = 5(2) + 11$$

Then at each step we rewrite the formula as we solve it, so

$$Y' = 10 + 11$$

then after adding the numbers we find

$$Y' = 21$$

It might be tempting to skip rewriting steps in simple formulas, but that is a bad habit to get into because it will hurt you when solving the more complex formulas.

Proportions and Percentages

Throughout this book, I will talk about proportions and percentages. A **proportion** is a number between 0 and 1. It is the number of times something happens (the frequency, or f) divided by the number of times it could happen (N). Therefore, the formula for proportion is

$$p = \frac{f}{N}$$

For example, if I were to ask how many of the students in my class grew up outside of the state where I teach (Pennsylvania), out of 20 students maybe 9 will raise their hand. That means that the proportion of students in the class who grew up outside of Pennsylvania is .45 (9/20). A proportion is a number between 0 and 1 since it could be possible that none of my students grew up outside of Pennsylvania, making the proportion 0 (0/20), or all of them did, making the proportion 1 (20/20). A proportion is never expressed as a fraction.

A **percentage** is a proportion multiplied by 100:

$$\% = \frac{f}{N} \times 100$$

This means that a percentage will range between 0% and 100%. Using the above example, I could say that the 9 of my 20 students who grew up outside of Pennsylvania constituted 45% of my students in that particular class ($.45 \times 100 = 45\%$). It is important to note the difference, and we will revisit proportions in the chapter on probability.

Rounding

When performing calculations in statistics, it will be necessary to round your numbers. In this book, we round to two decimal places (to the nearest hundredth). To do this, we look at the thousandths place (the third number to the right of the decimal place). If this number is less than 5, then just drop it. If it is 5 or greater, round the second number after the decimal one place up (increase the hundredths digit by one) before dropping the other digits. If your rounding would result in a whole number, add two zeros in the decimal places:

10.94578994 rounds to 10.95
1104.044417 rounds to 1104.04
7.9967054 rounds to 8.00

MYTHS OF TAKING STATISTICS

As long as I have been teaching statistics I have encountered two consistent myths about this class. The first is from students who tell me, "I'm going to fail this class because I am bad at math." This is a myth for two reasons. One, as I have demonstrated, the math used in statistics is not complicated. Two, if you are not good at math you need to work hard to get better. Mathematical ability is not innate like being double-jointed

(some have it and some do not). Even if you think you are bad at math you will do well in statistics if you (1) are careful and (2) work hard. Do not get down on yourself.

The second myth I hear is when my students tell me, "I will do great in this class because I am great at math." There are two reasons why this is a myth. One, my experience is that students who are good at math get overconfident and make mistakes. For example, when squaring 4 they get 8 (4×2) instead of 16 (4×4). Two, as I stated earlier, the emphasis in statistics is knowing how to interpret the number. Getting the correct number when you solve for a formula is very important, but it is pointless if you do not know what that number means.

VARIANCE AND ERROR

Whenever a measurement is taken, we can expect there to be **variance**, which means differences. For example, if I were to ask Democrats and Republicans how they are going to vote in an upcoming election, it is fair to assume that I will get different answers. There will be variability in their responses, but that variance is due to their political orientations. Likewise, if I were to ask criminal justice majors and social work majors about their attitude toward the death penalty, it is conceivable that the two groups

will differ in their answers (let us assume that criminal justice majors would be more in favor of the death penalty, and social work majors would be more against). Again, I am taking two measures and getting differences, and the differences could be attributed to the difference in majors. In statistics, we refer to this as **true variance**.

However, let us say that a series of political polls are taken that show a Republican candidate for office getting 55% of the votes among likely voters. One day a poll is released that shows the same candidate who was getting 55% of the vote is now polling at 30%. Now, there is a difference (variance) between that poll and all of the others conducted before it, so it raises the question, "Why is there such a difference in this poll?"

We inquire into the methods used by this particular pollster and find that they oversampled Democrats so that they accounted for 70% of her sample. Since Democrats make up about half of the voting population, we would say that Democrats were overrepresented in this person's poll. So here we see variance (55%–30%), but the variance is due to a mistake made on the pollster's part. This is referred to as an inaccurate measurement or **error variance**. The polls should all be roughly the same (within the margin of error), but that is not the case if a pollster uses bad sampling techniques. Keep in mind as you make your way through this book that all measurements are subject to error. Statistical analysis allows us to state the amount of error (or conversely, the degree of accuracy) in our measurements.

POWER AND ROBUSTNESS

Statistics allow for us to determine if the differences we find are due to true variance or error variance. A statistic has **power** if it can differentiate between true variance and error variance, and a powerful statistic can even tell us how much variance there is. This is because the statistics we use all have assumptions

for their use. The people who created the statistics that we use did so to solve particular problems or answer specific questions. Therefore, there is not a one-size-fits-all statistic. The correct statistic to use is the one appropriate for a certain kind of data (which we will explore further in the next chapter). However, **robust** statistics are the kind that can tolerate some (emphasis on the word some) violations of its assumptions. For example, when we learn about the independent samples t-test we will see that one of its assumptions (think, rules for using it) is that we cannot have very different sample sizes of the two groups we are comparing. It would be ideal if both groups were the same exact size, but since the independent samples t-test is a robust test the numbers can be slightly different. However, they cannot be wildly different sample sizes as that would violate one of the assumptions of independent samples t-test.

CHAPTER SUMMARY

We use research and statistics to better understand the social world. Data analysis helps our understanding by analyzing the relationships between variables and creating hypotheses so we can test theories. Different types of statistics are used to analyze samples and populations. We also use different statistics depending on the purpose of our research. Choosing the right statistic is very important and will help us distinguish between variance and error. This book is concerned with the collection, organization, and analysis of data using established, well-defined procedures.

FORMULAS FOR CHAPTER 1

Proportion: $p = \dfrac{f}{N}$

Percentage: $\% = \dfrac{f}{N} \times 100$

1. For the following variables, state what the possible attributes are. (Hint: In order to make them exhaustive you may have to make one of the attributes "other.")
   a. College major
   b. Year of graduation from high school
   c. Favorite genre of book to read for pleasure
   d. Favorite type of music to listen to
   e. Religion

2. For the following relationships, state which is the independent variable and which is the dependent variable.
   a. College major and attitude toward drug legalization
   b. Gender and attitude toward the death penalty
   c. Annual salary and level of education
   d. Rate of juvenile delinquency in a neighborhood and the number of single-parent households
   e. Number of packs of cigarettes smoked and incidences of lung cancer

3. Round the following numbers to two decimal places:
   a. 22.141456
   b. 17.990123
   c. 77.32890
   d. 51.012121

4. Compute the following:
   a. $(9 + 8)(12 + 26) - (6^2 - 10)$
   b. $(5^2 + 3^2)(2^2 + 4^2)$
   c. $(5 + 5)(9 \div 3)$

5. What is the proportion of each of the following? (Round to two decimal places)
   a. 1/10
   b. 15/50
   c. 27/1000
   d. 22/500
   e. 17/40

6. Transform each answer in number 5 into a percentage.

7. Solve the following equation where $a = -2$ and $b = 5$:

$$(a + b)(a^2 + b^2)$$

8. Solve the following equation where $a = 7$ and $b = 9$:

$$\frac{a + b}{\sqrt{b}}$$

9. Solve the following equation where $a = 16$ and $b = 4$

$$\left(\frac{a - b}{b}\right)\sqrt{b}$$

10. There were 50 recruits at the police academy. Fifteen ran the mile test, 20 were in the weight room, 10 were taking a written exam, and 5 ate lunch.
    a. What proportion of the recruits ate lunch?
    b. What proportion were not taking a written exam?
    c. What percent ran the mile test?

CHAPTER 2

UNDERSTANDING THE DATA: VARIABLES AND LEVELS OF MEASUREMENT

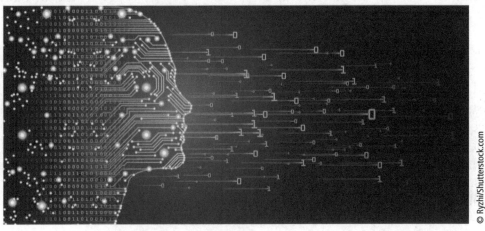

© Ryzhi/Shutterstock.com

We think like researchers. Although we might not use the same terms they use, the thinking we employ is a lot like the thinking researchers employ. For example, do you like to people-watch? If you are like a lot of the people I know, you like to sit in a crowded area (perhaps a quad on campus, in the dining hall, or at a shopping mall), and just watch the people walk by. If someone you knew were to approach you and ask what you were doing, you would answer, "people watching."

In research, we call this collecting data. All of the observations we make are stored in the wonderful supercomputer we call our brain. (Although comedian Emo Philips once said, "I used to think my brain was my most wonderful organ. Then I realized who was telling me this.") And since the data are being collected via one or more of our five senses, we would say that people watching is collecting **empirical** data. The word empirical means, according to the Merriam-Webster dictionary, "Originating in or based on observation or experience," so think of empirical data as data that are "in the world." Now, the next time you are people watching and someone asks you what you are doing, I am not suggesting that you answer, "Empirically collecting data," but you get my point.

While I believe that we think like researchers, I also believe that we talk like statisticians. For example, years ago when I turned 40 most of my friends were all turning 40 around the same time. One of my friends who was a very good runner in high school (cross-country and indoor and outdoor track) told me that for his 40th birthday (which was 3 months away) he was going to run a marathon. However, he had not run in the previous 20 or so years (unless, presumably, he was being chased), and over that same time period he had smoked at least one pack of cigarettes a day.

These are not prerequisites for running a marathon.

So when he called me and told me about his goal of running a marathon for his 40th birthday and asked me, "What do you think?" I had to be honest. So, I told him that I thought he had a better chance of getting pregnant in 3 months than running a marathon, but that it was a worthwhile goal and I wished him the best.

Well, my friend was less than happy with my forthrightness, and he called me once a week during his training to give me updates on his progress. He would call and say, "I averaged 5 miles a run this week," or, "I totaled 23 miles this week."

Now if you think about what he said, he was speaking statistically. The words "average" and "total" are words we use in statistics. Moreover, he was speaking statistically to **summarize** his experience. Rather than call me every day that he ran, he only called once a week and summarized his training in terms of an average or total. We do this all the time.

We also speak statistically when we **generalize** and **predict**. If you have ever said, "We can expect a lot of snow this winter," what you are doing is making a statement based on your past experience (or stored data). Perhaps over the previous few winters there was a lot of snow, so your generalization makes sense. Also, if you have ever said, "The earlier you start studying for that statistics exam, the better you will do," you are speaking statistically to make a prediction. The prediction is about the relationship between two variables—the independent variable, "time spent studying," and the dependent variable, "grade on an exam." Whenever we summarize, generalize, or predict, we are speaking like statisticians.

MEASUREMENT

In research, measurement starts with a concept. **Concepts** are words we use to represent mental images. For example, recidivism is a concept. **Conceptualization** is the process of defining what the concept represents. Therefore to conceptualize recidivism, we would say that it is being arrested for an offense after being previously arrested (for the same or a different offense).

This process of conceptualization makes it possible to then measure the concept. **Measurement** is the process of applying numbers to a concept in order to represent its properties. Think of measurement as scoring. Once we have a clear definition of recidivism, we can then count the amount of recidivism in different areas so that we can compare different recidivism rates.

These concepts are also referred to as **units of analysis**. A unit of analysis is who or what we study. So when we take a sample, units of analysis are what make up that sample. The most common units of analysis in Behavioral Sciences research are individuals, groups, organizations, and social artifacts. With individuals, we could take a sample of nursing majors and ask them their gender, age, race, and so on. This would give us a composite of what a typical (though not all) nursing major was like.

With groups, we could compare the average NCLEX-RN (National Council Licensure Examination-Registered Nurse) scores for nursing graduates of different colleges or universities. In this instance, we are not examining individual nursing majors but groups of nursing majors. In examining organizations, we could compare the recidivism rates of different treatment facilities for juvenile delinquents (comparing different organizations to each other). And social artifacts are the products of human behavior. So if we were to measure the amount of violence in different television shows, we would be examining a social artifact (the television shows).

QUANTITATIVE AND QUALITATIVE VARIABLES

Some of the variables we examine are **quantitative,** and some are **qualitative.** With quantitative variables, the number we use to represent something has a direct meaning. For example, when someone says that they are 22 years old, we know what that number represents. If a person is 5'10" tall, that number does not need to be explained. If on an exam worth 50 points one student earned 45 points and another earned 35 points, we know how they did relative to each other (which one got a higher score on the exam). Another characteristic of quantitative variables is that they differ in amount. Age, height, and scores on an exam are examples of quantitative variables.

Data Collection Tools

Quantitative
 Online-Web
 Face-to-Face
 Phone
 FedEx Mail
 Central Location Intercept

Qualitative
 Online Forums
 Online Communities
 Web Survey
 Groups-Triads-Dyads
 Depth Interviews [IDIs]

© arka38/Shutterstock.com

Qualitative variables differ in kind rather than amount. With a qualitative variable we are putting units of analysis into different categories (which is why qualitative variables are also called categorical variables). For example, political affiliation is a qualitative variable because people belong to different political parties, but the different parties do not indicate the amount of anything. Gender, eye color, and zip code are examples of qualitative variables. Zip codes are a number, but that number does not represent the amount of anything.

RELATIONSHIPS

We collect data to determine if there is a relationship between variables. For example, does the amount of time a student studies have any effect on their exam grade? To answer this question, we would collect data on a student's score on an examination and then ask them how many hours they studied for that exam (let us assume that they are honest). We would do this a number of times and put the data in a chart like the one in Table 2.1.

TABLE 2.1 Hours Studied and Score on Examination

Number of Hours Studied (X)	Score on Exam (Out of 100 points) (Y)
0	35
0	35
1	41
2	45
3	75
4	85
4	86
4	88
5	87
6	95

What we do is look for a pattern among the data. If we can discern a pattern, then there is a relationship between the variables. If the relationship is consistent, then it is a strong relationship. Consistent means that as one variable changes (in this case, number of hours studied), the other variable (score on the exam) changes as well. As can be seen from Table 2.1, it is apparent that the students who studied more did better on the exam than students who studied for fewer hours. If we were to predict the relationship, it could be summed up as: the more you study, the better you will do on exams.

However, this relationship is not a perfect relationship. To have a perfect relationship, you would see one value of the X variable and one and only one value for the Y variable.

As can be seen from Table 2.1, there are three different scores for students who studied for 4 hours (85, 86, and 88). The strength of a relationship varies due to the relative consistency of the variables. In Table 2.2, we see a less consistent, hence weaker, relationship.

TABLE 2.2 Hours Studied and Score on Quiz

Number of Hours Studied (X)	Score on Quiz (Out of 25 Points) (Y)
0	10
0	18
1	18
2	21
3	8
4	20
4	19
4	12
5	22
6	21

As can be seen from Table 2.2, it is hard to discern the relationship between hours of study and score on a quiz because of the inconsistency. For example, even though there are three different scores associated with 4 hours of study, just as there were in Table 2.1, in Table 2.2 there is a greater spread in these three scores (20–12) than there is in Table 2.1 (88–85). We would say that for this sample the relationship between hours of study and score on a quiz is weaker (less consistent) than the sample in Table 2.1.

EXPERIMENTS AND CORRELATIONAL STUDIES

In research we want to determine if there is a relationship between variables and, if so, how strong that relationship is. However, we have different ways of making that determination. How a study is designed is how it is conducted; how large of a sample size, how data are collected, how units of analysis are selected, and so on, are all part of a study's design. This is important because different designs require different procedures for studying relationships.

Experiments

Basically, experimentation involves (1) making an observation, (2) taking some kind of action, and (3) observing the consequences of that action. For example, I get in my car to drive to work and it does not start (observation). I think that my battery is dead, so I get one of my neighbors to give me a jump-start (action). Then I try to start my car (observing the consequences). If my car starts, then the experiment was a success. If not, then I have to try something else. (My automotive expertise is extremely limited, so my next action will be to call AAA—the American Automobile Association—and their road-side assistance program.)

There are three parts to an experiment. The first part is the "independent and dependent variables." In an experiment, the independent variable is also referred to as the "treatment variable" or the "stimulus." As discussed in Chapter 1, the dependent variable is the effect (sometimes called the "outcome variable") in an experiment.

The second part of an experiment is "pre and post testing." Subjects are measured on a dependent variable, exposed to a stimulus (the independent variable), and then remeasured. Any differences noted between the first and second measures on the dependent variable are then attributed to the influence of the independent variable. The "pretest" is the measure of the dependent variable prior to the introduction of the stimulus, and the "posttest" is the measurement of the dependent variable after the independent variable has been introduced.

The third part of an experiment is the "experimental and control groups." The "experimental group" is the group to which the independent variable has been administered, and the "control group" does not get the independent variable. Using a control group enables the researcher to control for the effects of the experiment. If after administering the independent variable there is a change in the experimental group and not the control group, then any differences can be attributed to the independent variable. Two key aspects to conducting an experiment are (1) the manipulation of the independent variable (giving it to one group and not the other), and (2) controlling the aspects of the experimental situation.

For example, let us say we wanted to test whether listening to music has an effect on test scores. We randomly assign participants to two groups and give each a pretest measure. The control group then takes a posttest without listening to music, while the experimental group takes the posttest with music playing. Then we compare the experimental group's pretest and posttest scores, and also the experimental group's posttest scores with the control group's posttest scores. Any differences we find could be attributed to music being played while the experimental group was taking their posttest.

Correlational Studies

Sometimes it is not possible to conduct an experiment, so we conduct correlational studies instead. With a correlational study, we measure participant's scores on two variables and then determine whether there is a relationship between the two. Unlike an experiment where an independent variable is manipulated (or changed), in a correlational study the researcher does not manipulate anything.

For example, it is well documented that the higher one's educational level, the more money they are going to make. This is known because data have been collected on the independent variable (level of education) and the dependent variable (annual income).

However, it would be impossible to conduct an experiment to test this relationship. You would have to randomly assign individuals to different groups and manipulate their level of education. Since that cannot be done (you cannot make some people drop out of high school, make others graduate from high school but not go to college, make some graduate from college, etc. . .), all we can say is that there is a correlation between these variables.

Causality

Just because we have a correlation (relationship) between variables does not mean that we can say one causes the other. If we can establish a correlation between hours spent studying and grade on an examination (e.g., students who spend more time studying earn higher exam scores), this does not mean that we can say that studying longer will cause you to get a higher score. This is because correlations do not account for any intervening variables (discussed in Chapter 1) that could effect this relationship. Intelligence, motivation, and amount of sleep the night before the exam (among other factors) could all have an effect on a students' grade. Correlation does not mean causation.

LEVELS OF MEASUREMENT

Numbers only make sense when they are placed in the proper context. On a questionnaire the number 1 that corresponds to a persons' religion is different than the number 1 found in the score in a soccer game. In other words, not all numbers can be treated equally. The information that a number will convey depends on the level of measurement, and there are four such levels.

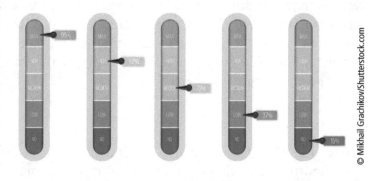

© Mikhail Grachikov/Shutterstock.com

There are two main reasons why understanding the type of data we have is important. One is that it will help us understand statistics on a conceptual level. As stated in chapter 1, whether you will ever conduct statistical analysis after this class is unknown, but you will work with statistics. The other reason why this is important is that we cannot conduct any statistical analysis without first understanding what we are measuring, which means understanding the type of data we have.

For example, if I asked you to measure someone's weight you would use a scale, you would not pick them up in your arms. But if I were to ask you to measure the temperature outside you would not use a scale, you would get a thermometer. One of the reasons mistakes are made in statistics (error variance) is that people sometimes choose an instrument (statistical tool) that is wrong for what they are measuring. Choosing the wrong analytical tool is like measuring the temperature outside with a bathroom scale. A scale can be useful for taking measurements, just not for temperature.

Nominal

The numbers assigned to **nominal**-level variables do not indicate an amount, instead they are used to signify different categories. Republicans, Democrats, Libertarians, Greens, etc. . ., differ from each other in kind, not in any quantitative way. The numbers we attach to those attributes (for the variable, political party) have no meaning other than to distinguish (or categorize) one party from another. Therefore, political party is nominal-level data. Other examples of nominal-level data are the numbers on basketball jerseys or zip codes, which are used to differentiate one player or one area from another, respectively. While these are numbers (whereas political parties are not), remember that these numbers do not mean any quantitative difference. The differences between different jersey numbers and zip codes are categorical only. Nominal-level variables indicate that one unit of analysis is qualitatively different from another. The word nominal comes from the Latin *nominalis* meaning pertaining to name. So if we are conducting a study and an independent variable is sex, then we could classify males as 1 and females as 2. Since the numbers are used only to put participants into one category or another, we could pick any number we want, say, assign males as 22 and females 201,104, or males as 1,000,988 and females 17. As long as different numbers are used to indicate different categories, the numbers themselves are irrelevant. Another important characteristic of nominal-level data is that it is not ranked. Your own political leanings notwithstanding, one political party is not objectively better than another, and so the numbers assigned to them do not indicate superiority or more in amount, just like we could not say that a player wearing jersey number 22 is twice as good as a player wearing number 11, or one-third as good as a player wearing number 66.

Ordinal

Ordinal-level variables are also qualitative; so the numbers we assign to the categories are not important since they do not have a direct meaning or measure the amount of something. However, if a variable can be arranged in a meaningful way, or order, then it is ordinal-level data. The numbers used in ordinal-level data indicate a hierarchy of levels for the variable under consideration. Ordinal comes from the Latin *ordinalis* meaning pertaining to order. Rankings in a police department form an ordinal-level variable because some officers have a higher rank than others. For example, the Cape May, New Jersey, police rankings are as follows: Patrolman, Sergeant, Lieutenant, Captain, and Chief. We could code the rankings as 1 for Patrolman, 2 for Sergeant, up through 5 for Chief. These numbers do not indicate an amount and are used only to classify the different rankings. Therefore, we could use any numbers for the rankings: 04 for Patrolman, 57 for Sergeant, 58 for Lieutenant, 122 for Captain, 5,987 for Chief. While the numbers themselves are not important, the order of the numbers is very important. The higher numbers have to indicate a higher rank. It would be incorrect to code Patrolman as 57 and Sergeant as 04 since Sergeant is a higher rank than Patrolman. Another characteristic of ordinal-level variables is that the differences between the rankings are not equivalent; we have no way of discerning the degree of difference between the values. In other words, we are not always sure how much difference exists between the categories. The time it takes for a Patrolman to become a Sergeant is not the same as the time it takes for a Lieutenant to become a Captain. In addition, there are other factors involved (e.g., passing an examination to get a promotion or having a college degree). So, similar to nominal-level variables, the numbers used for ordinal-level data are not important, but unlike nominal-level data, the order of those numbers is very important.

Interval

The next level is **interval**-level data. Unlike nominal and ordinal data, interval-level variables are quantitative—the numbers do indicate an amount. In addition, the differences between interval scores are equivalent (interval comes from the Latin *intervallum* meaning the space between). Score on an examiation is an example of interval-level data. Someone who scores 90 on an exam did better than someone who scored an 80. Additionally, the difference between the scores of 80 and 90 is the same as the difference between the scores of 60 and 70. A key distinction of interval-level data is that there is not a true zero. There are scores of 0, but a true 0 means the absence of something. Someone who earns a 0 on an exam did get a score, but they did terribly; it does not mean that they did not get a score. Another example of interval-level data is temperature (e.g., in Fahrenheit). Higher numbers indicate higher temperatures, and the difference between 55 and 60 degrees is the same as the difference as 65 and 70 degrees. Also, a temperature of 0 degrees does not indicate the absence of weather, it just means that it is really cold.

Ratio

The last level of measurement is **ratio**-level data. Like interval-level data, ratio variables indicate the amount of something, and the differences between units are equivalent. However, the differences between ratio- and interval-level data are that with ratio data there is a true zero (the absence of something), and also that we can have meaningful ratios. Meaningful ratios are ratios that make sense. For example, when I started my career in juvenile probation I was assigned 0 cases for the first three months that I was training. This means I literally had no cases on my caseload. In my fourth month, I was assigned 5 cases, and by my fifth month I had 10 cases. So in my fifth month I had twice as many cases as I had in my fourth month. Or, in my fourth month I had half as many cases as I had in my fifth month. It is awkward to say that 80 degrees is twice as hot as 40 degrees because there is no true zero point, so interval-level data do not lend themselves to meaningful ratios.

As you can see from Table 2.3, each level of measurement has the same characteristics as the preceding levels. For example, nominal-level data only categorize the attributes. Ordinal-level data categorize the attributes (like nominal-level data) and also rank them. Interval-level data categorize, rank, and are characterized by an equal distance between the attributes. Finally, ratio-level data categorize, rank, and have an equal distance between attributes, and also have a true zero.

TABLE 2.3 Characteristics of the Levels of Measurement

Level of Measurement	Categorizes	Ranks	Equal Distance between Attributes	True Zero
Nominal	Yes	No	No	No
Ordinal	Yes	Yes	No	No
Interval	Yes	Yes	Yes	No
Ratio	Yes	Yes	Yes	Yes

DISCRETE AND CONTINUOUS VARIABLES

Variables can be either discrete or continuous. A **discrete** variable is one that does not allow for fractions or decimals—they must be measured in whole amounts. Discrete variables only take on a finite number of values. So, year in college (freshman, sophomore, etc...), number of grammatical mistakes in a term paper, and number of car accidents in a month are discrete variables because you cannot be in two college years at the same time, make 1/2 a mistake in a term paper, or be involved in 7.47 accidents.

© Sharaf Maksumov/Shutterstock.com

Continuous variables, on the other hand, can be measured in fractions or decimals and can take on (potentially) any numerical value. For example, if you had $10.25 in your pocket, we could say that you have 10 and .25 dollars. When criminals get sentenced, sometimes they get 11½ to 23 months (you can have half of a month). Your grade point average can be 3.0, 3.2, 3.267, etc... This does not mean that we have to use decimals or fractions when measuring a continuous variable, just that we have the ability to, which we do not have with discrete variables.

CHAPTER SUMMARY

In research, we conduct experiments and correlational studies to better understand the relationship between variables. We judge the strength of a relationship by how consistent it is. In statistics, some of the variables have a numerical value that measures the amount of something, while other variables are assigned a numerical value that does not have a direct meaning. Understanding the different levels of measurement is the first step in understanding the kind of data we have.

1. Which of the following three relationships is the strongest? Which is the weakest?

TABLE 2.4 Comparison of Parent and Children Heights

Average Height of Parents (in inches)	Average Height of Children (in inches)
62	60
65	66
68	66
69	70
70	70
72	72
74	73
74	75
75	78
75	78

TABLE 2.5 Comparison of Milk Consumed and Cavities

Number of Glasses of Milk Consumed Per Day	Number of Cavities
0	3
0	3
1	2
1	2
1	2
2	1
2	1
3	0
3	0
3	0

TABLE 2.6 Comparison of Carrots Consumed and Examination Scores

Number of Carrots Consumed Per Day	Score on a Statistics Examination (Out of 100)
0	97
0	27
1	100
1	80
2	78
2	56
2	95
3	33
4	90
5	76

2. For the following, state what the unit of analysis is. (Hint: it will be either individual, group, organization, or social artifact.)
    a. Female Criminology majors _____
    b. Men shopping at a mall _____
    c. The effectiveness of a police department in reducing crime _____
    d. Cartoons depicting violent acts _____

3. State which of the following scenarios would best be tested using an experiment or a correlational design:
    a. The effect of growing up in a single-parent household on juvenile delinquency _____
    b. The effect of having a parent who is a nurse on choice of college major _____
    c. The effect of the amount of light in a room while studying on test scores _____
    d. The effect of having at least one pet while growing up on whether someone is a vegetarian _____
    e. The effect of a memory drug on Alzheimers patients _____
    f. The effect of room temperature on sleep _____
    g. The effect of number of absences from class on grade _____
    h. The effect of temperature in a city and the amount of ice cream sold _____

4. Complete the following table with the correct characteristics of each variable:

Variable	Qualitative or Quantitative?	Discrete or Continuous?	Level of Measurement
College Major			
Age			
Political Party			
Military Rank			
Number of Days Sick from Work			
Weight			
Level of Education (High School Diploma, Associates Degree, etc. . .			
Favorite Color			
Number of Traffic Accidents in a Given Month			
Number of Prior Offenses			
Telephone Numbers			
Hair Color			
Year of High School Graduation			
Annual Income			

CHAPTER 3

FREQUENCY DISTRIBUTIONS AND GRAPHS

© PixMarket/Shutterstock.com

INTRODUCTION

Chapter 1 stated that a purpose of descriptive research is to paint a picture. We use descriptive statistics to create an image of data. To do this, we need to know certain things about the data. A frequency distribution is a count of how frequently each attribute of a variable occurs. For example, the number of college sophomores in a class, the number of sick days from work, or the number of prior arrests among a group of criminals. As statisticians, our job is to organize and summarize data. Frequency distributions not only help us do that, but after data are organized into a frequency distribution we can extract more information from the data, which would be very difficult to do if the data were left in its raw form.

All data comes to us as raw data. Let us say you asked 35 people what their favorite color is. You might get the following results:

Red, blue, blue, brown, black, red, black, green, pink, pink, red, green, blue, orange, blue, red, black, blue, green, red, yellow, red, blue, green, pink, yellow, purple, red, green, black, brown, purple, pink, green, and red.

The reason this is called raw data is that it is unorganized. However, when we transform this data into a frequency distribution, we organize it and make it easier to understand. For example, how many people said that pink is their favorite color? To answer this question, you would go back through all of the colors and pick out the number of pinks. This might take just a few seconds, but see how much faster it is to answer that question once the data are organized.

TABLE 3.1 Favorite Colors

Color	Frequency
Red	8
Blue	6
Brown	2
Black	4
Green	6
Pink	4
Orange	1
Yellow	2
Purple	2

This frequency distribution makes the information more compact and comprehensible. Since this is nominal-level data, the order of the colors is not important. This data could have been organized in alphabetical order or by the frequencies (lowest to highest, or highest to lowest). Now that the raw data are in a frequency distribution, you can see how much neater and organized it is than in its raw form. However, this frequency distribution shows a limitation of nominal-level data—there is not much you can do with it, statistically speaking. With interval- or ratio-level data, we can create much more detailed frequency distributions.

SIMPLE FREQUENCY DISTRIBUTIONS

A simple frequency distribution is the most common way to organize and present data. The symbol f stands for frequency, which is the number of times an attribute occurred. With interval- and ratio-level data, we will do much more than simply display how often something happened.

Instead of favorite color, let us organize the following examination scores (out of 100):

90	95	80	80	81	82	92	93	93	95	96	85	85
85	90	90	88	88	90	90	96	95	84	84	90	90

This data in its raw form provides information about individual performance, but not the overall collective performance. In creating our simple frequency distribution, we start out by ordering the scores from lowest to highest. Since the range of our scores goes from a low of 80 to a high of 96, we include all values between 80 and 96. Even if no one had a score in that range, we still include all of the values; we never skip numbers in a frequency distribution. Then we add a column (*f*) that indicates how often each score occurred. If no one has a certain score, we do not skip it. Instead, we include it in the **Score** column and give that score a frequency of 0 (Table 3.2).

Now this is a start, but we can do more with higher levels of measurement. This means we will add more columns to extract more information from the data.

The next column we will add is cumulative frequency (*cf*), which is the number of scores so far. This is valuable when we are not only interested in the frequencies for a certain score but also how many scores are above or below a certain value. Then we will calculate the proportion for each score and add a column for that (*p*). (Remember from Chapter 1 that the formula for proportion is *f/N*). The next column is cumulative proportion (*cp*), which is the total proportion so far (or, the proportions above or below a certain value). Then the last two columns are percentage (**%**) and cumulative percentage (*c***%**), which is also referred to as a score's percentile.

To add the cumulative frequency column, we start with the frequency of the lowest score. The first cumulative frequency value is the same as the first frequency value. Next we take that first cumulative frequency value (in this case 2) and add it to the frequency in the next row (in this case 1) to get a *cf* of 3 for the score 81. We take that *cf* value, add it to the next frequency, and so on (adding across columns). We do that all the way down until the total cumulative frequency is the same number as our sample size (Table 3.3).

TABLE 3.2 Frequency Distribution

Score	f
80	2
81	1
82	1
83	0
84	2
85	3
86	0
87	0
88	2
89	0
90	7
91	0
92	1
93	2
94	0
95	3
96	2

TABLE 3.3 Frequency Distribution

Score	f	cf
80	2	2
81	1	3
82	1	4
83	0	4
84	2	6
85	3	9
86	0	9
87	0	9
88	2	11
89	0	11
90	7	18
91	0	18
92	1	19
93	2	21
94	0	21
95	3	24
96	2	26

There are two ways to quality control yourself (make sure you are not making any mistakes). One is to add up the values in the frequency column before you complete the cumulative frequency column—this number should be the same as the sample size. The other way is to complete the cumulative frequency column, and the total cumulative frequency should be the same as the sample size. (In this case it is. The total *cf* is 26, and we have 26 scores.) If this is not the case, then go back and see where you went wrong. It is much easier to fix a simple frequency distribution early on than when it is all done.

For the next column, proportion (**p**), we take the frequency from the second column and divide it by the sample size (in this case, 26). We do this for every frequency, even if the frequency is 0. When we do this, we round to two decimal places (as discussed in Chapter 1). Proportions are important because this is how we calculate percentages, which is helpful because it is usually easier to interpret that 27% of the students scored 90 than it is to say that 7 out of 26 students scored 90 (Table 3.4).

The next column is cumulative proportion (**cp**). This works the same way as cumulative frequency. However, instead of saying how many frequencies there are so far, it says how many proportions there are so far. The first value for **cp** is the same value as the first proportion. We start there, then add proportions across columns (Table 3.5).

TABLE 3.4 Frequency Distributions

Score	f	cf	p
80	2	2	.08
81	1	3	.04
82	1	4	.04
83	0	4	.00
84	2	6	.08
85	3	9	.12
86	0	9	.00
87	0	9	.00
88	2	11	.08
89	0	11	.00
90	7	18	.27
91	0	18	.00
92	1	19	.04
93	2	21	.08
94	0	21	.00
95	3	24	.12
96	2	26	.08

TABLE 3.5 Frequency Distribution

Score	f	cf	p	cp
80	2	2	.08	.08
81	1	3	.04	.12
82	1	4	.04	.16
83	0	4	.00	.16
84	2	6	.08	.24
85	3	9	.12	.36
86	0	9	.00	.36
87	0	9	.00	.36
88	2	11	.08	.44
89	0	11	.00	.44
90	7	18	.27	.71
91	0	18	.00	.71
92	1	19	.04	.75
93	2	21	.08	.83
94	0	21	.00	.83
95	3	24	.12	.95
96	2	26	.08	1.03

You can see that the total cumulative proportion is greater than 1, even though a proportion is a number between 0 and 1. This is due to the fact that we rounded the proportions. We do this to make a clean distribution. The proportion for a score of 80 is actually .076923007 (2/26). No one wants a column in a frequency distribution to be 9 digits wide, so we round to two decimal places.

There are two more columns. We want to know the percentage of a certain score, so that is our next column (%). As you recall from Chapter 1, a percentage is a proportion multiplied by 100 ($f/N*100$). We can do that or just move the decimal place in the proportion column two spaces to the right (Table 3.6).

TABLE 3.6 Frequency Distribution

Score	*f*	*cf*	*p*	*cp*	%
80	2	2	.08	.08	8
81	1	3	.04	.12	4
82	1	4	.04	.16	4
83	0	4	.00	.16	0
84	2	6	.08	.24	8
85	3	9	.12	.36	12
86	0	9	.00	.36	0
87	0	9	.00	.36	0
88	2	11	.08	.44	8
89	0	11	.00	.44	0
90	7	18	.27	.71	27
91	0	18	.00	.71	0
92	1	19	.04	.75	4
93	2	21	.08	.83	8
94	0	21	.00	.83	0
95	3	24	.12	.95	12
96	2	26	.08	1.03	8

The last column is cumulative percent (*c%*), which is also known as a score's percentile. A percentile is the percentage of scores at or below a certain value. If we say that someone scored at the 90th percentile on a test, it means that they did as well as or better than 90% of the people taking that test. Notice that we do not know their raw score, but we do not have to. A percentile references a score *vis-à-vis* everyone else taking that test. To calculate the cumulative percentage, we can multiply the cumulative proportion by 100 (*c%* = *cp**100), or move the decimal point in the *cp* column two places to the right.

Now the simple frequency distribution is complete. Not only are the data better organized than in their raw form, but we can extract more information about the data. As stated earlier, whether you will calculate statistics beyond this class is unknown, but you will have to work with statistics, and that includes interpreting simple frequency distributions. Looking at Table 3.7, we can answer questions about the data that would have been difficult, if not impossible, had it been left in its raw form.

TABLE 3.7 Frequency Distribution

Score	f	cf	p	cp	%	c%
80	2	2	.08	.08	8	8
81	1	3	.04	.12	4	12
82	1	4	.04	.16	4	16
83	0	4	.00	.16	0	16
84	2	6	.08	.24	8	24
85	3	9	.12	.36	12	36
86	0	9	.00	.36	0	36
87	0	9	.00	.36	0	36
88	2	11	.08	.44	8	44
89	0	11	.00	.44	0	44
90	7	18	.27	.71	27	71
91	0	18	.00	.71	0	71
92	1	19	.04	.75	4	75
93	2	21	.08	.83	8	83
94	0	21	.00	.83	0	83
95	3	24	.12	.95	12	95
96	2	26	.08	1.03	8	103

To answer "how many people scored 95 on the exam?" we find the score of 95 and look at the number in the frequency column (3). If we ask "how many people scored 88 and below?" we find the score of 88 and then look in the cumulative frequency column (11). That would have been very difficult to answer had we not put the raw data into a simple frequency distribution. We know that .04 of the people scored a 92 on the exam by the corresponding number in the proportion column, and that .24 of the people scored 84 and below from the corresponding number in the cumulative proportion column. Finally, the answer to "what percent of the people scored 90 on the exam?" is answered by the corresponding number in the percentage column (27), and we know that a score of 93 is at the 83rd percentile by finding the corresponding number in the cumulative percent column. Imagine trying to ascertain this information about the data if it were left in its raw form.

GROUPED FREQUENCY DISTRIBUTIONS

Sometimes we have a large range of data that will not fit neatly into a simple frequency distribution. The data used for the simple frequency distribution had a range of 17, but what if we have a range of 49, like this:

40	49	50	43	45	46	60	88	89	63	85	67	76
67	55	44	78	89	60	50	51	78	54	68	71	72
73	56	58	77	77	56	55	66	54	54	78	88	88
81	81	89	88	82	76	76	75	78	89	66	67	80

TABLE 3.8 Grouped Frequency Distribution

Score	f	cf	p	cp	%	$c\%$
40–49	6	6	.12	.12	12	12
50–59	11	17	.21	.33	21	33
60–69	9	26	.17	.50	17	50
70–79	13	39	.25	.75	25	75
80–89	13	52	.25	1.00	25	100

With a wider range of data, it is better, for the sake of presentation, to group the values. We will use the same columns as a simple frequency distribution, but instead of one value under the **Score** column we will have a range of scores (called intervals).

If you are going to make a grouped frequency distribution, there are rules that must be followed. First, all of the intervals of scores must be equal (in this case the range is 10). It would be sloppy to have a first interval of 40–49, and the second interval be 40–54, then the third 55–70, and so on.

Second, the intervals must be mutually exclusive, which means that the scores must be able to fit into one and only one interval. You can't have the first interval be 40–49 and the second 49–58, because if someone scored 49 into which interval would you place them? And third, the ranges must be exhaustive, which means that you have to be able to account for every score in the data. Our range of 40–89 encompasses all of the scores.

From the grouped frequency distribution, we can see that 11 people scored between 50 and 59 and that 12% scored between 40 and 49. However, we lose some precision when we group our data; while we can see that 13 people scored between 80 and 89, we cannot tell how many people had a score of 83 (no one did).

GRAPHS

Graphs are another way to present data that demonstrate a relationship between an attribute and the frequency of that attribute. Sometimes, pictoral presentations of data can reveal information about the data that are not easily seen from a table. The type of graph we use depends on the type of data that we have, or the characteristics of the variable being measured. A graph will show the scores or attributes on the x-axis (also called the "abscissa") and the frequency on the y-axis (also called the "ordinate").

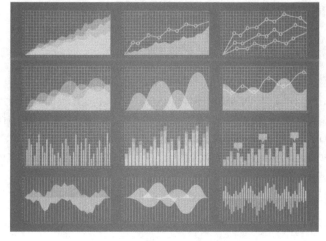

Pie Chart

A pie chart can be used with any level of measurement. The different slices of the pie indicate the attributes of the variable being examined. Pie charts are better suited for variables with few attributes, and a general rule is that a pie chart should not be used if a variable has more than six attributes.

Figure 3.1 shows a pie chart for the subject "college major." One of the problems with pie charts is that we have to estimate the percentages in each attribute. It is clear that there are slightly more criminal justice majors than nursing majors, but the exact numbers are unknown.

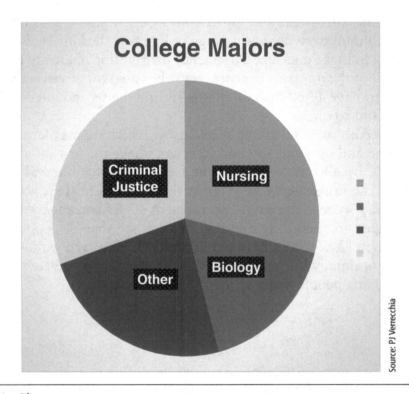

FIGURE 3.1 Pie Chart

Bar Graph

A bar graph is used when we have nominal- or ordinal-level data. Nominal-level data is categorical; ordinal-level data is also categorical, but implies ranked order. In a bar graph, a vertical bar is centered over each attribute on the x-axis and the adjacent bars do not touch. It is important that the bars do not touch because we want to show that there is not continuity between the attributes. Figure 3.2 shows a bar graph for the variable "year in school." Since this is ordinal-level data the bars do not touch (it is not continuous), but the order of the bars is important. The lowest rank (freshman) has to come first, followed by the next highest (sophomore), and so on.

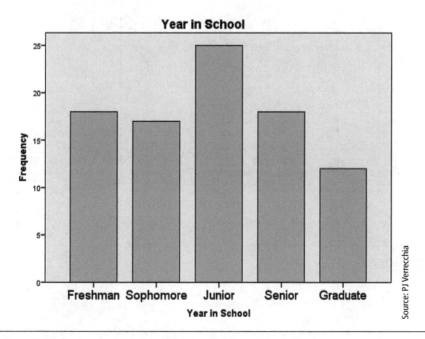

FIGURE 3.2 Bar Graph

Histogram

A histogram is used when we have interval- or ratio-level data, but a small range of data. A histogram is similar to a bar graph, but in a histogram the adjacent bars do touch. By having no spaces or gaps between the bars, we imply that there is continuity in the data. Figure 3.3 shows a histogram for age, which is a continuous variable, and so the bars touch.

Polygon

We use a polygon when we have a large range of interval- or ratio-level data. Though there is no set rule for what constitutes a "large range," a histogram with a range of 40 values would be too wide to look good, so we replace each bar with a single dot. This conveys the same information but with much less space. Then we connect each point with a line, as seen in Figure 3.4.

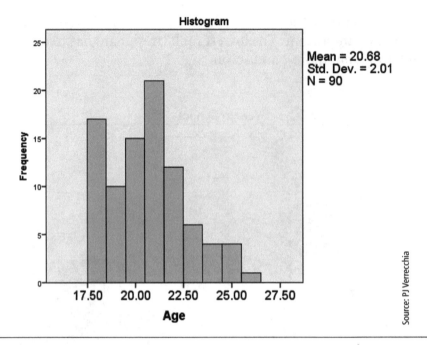

Source: PJ Verrecchia

FIGURE 3.3   Histogram

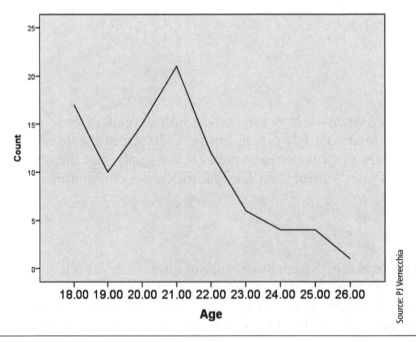

Source: PJ Verrecchia

FIGURE 3.4   Polygon

THE NORMAL DISTRIBUTION

The normal distribution (or the normal curve) is something that will be referred to throughout the rest of this book. It is a polygon that has a bell shape, indicating that data are normally distributed. The normal distribution is a theoretical distribution because it is not based on actual data but rather on how we assume data appear. An empirical distribution is based on a finite number of observations, while a theoretical distribution is based on an infinite number of observations. The normal curve is probably the most important distribution

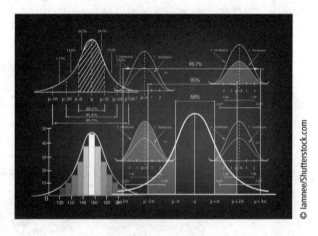

© Iamnee/Shutterstock.com

in statistics. This will become clearer as we explore the properties of the normal curve. (I will use the terms "normal distribution" and "normal curve" interchangeably.) There is a formula for the height of the normal curve, but it is beyond what you need for an introductory statistics course.[1]

The characteristics of the normal curve are as follows:

- It is symmetrical, which means that it is even on both sides. If you could fold the normal curve in half, the lines would match up perfectly;
- The scores with the highest frequencies are in the middle (this is the peak of the distribution);
- As we move away from the center of the distribution (the scores increase and decrease), the frequencies decrease. Even as the scores (or values) are increasing (to the right of the center), the frequencies (the number of times each score or value occurred) decrease;
- The scores with the highest and lowest values have the lowest frequencies; they are at the tail regions of the normal distribution.

The normal curve is a theoretical distribution, but it makes sense. For example, the average weight for men in the United States is 180 lb.[2] What this means is that if we could somehow weigh every man in the U.S. and graph it, the weight of 180 lb would be in the middle of the *x*-axis and it would have the highest frequency. As we move away from 180 lb (to 179, 181), then the frequencies would slightly decrease. The further we

---

[1] But if you really must know it is $\dfrac{1}{\sigma\sqrt{2\pi}}\,e^{-(x-\mu)^2/(2\sigma^2)}$

where μ is the mean, σ is the standard deviation, π is the constant 3.14159, and e is the base of natural logarithms and is equal to 2.718282. x can take any value from – infinity to + infinity.

[2] According to halls.md.

move away (to 130, 230), the more the frequencies decrease. Out in the tail regions, we would find the most extreme weights, but they would have the lowest frequencies. Since this a theoretical distribution, the tails are "asymptotic," which means that they never touch the *x*-axis.

The normal curve is very important in statistics because when we measure things in the world, many times, the scores will distribute themselves normally. Most of the scores will be in the middle, and the higher and lower values, occurring less frequently, will be in the tail regions. Figure 3.5 is a representation of the normal distribution.

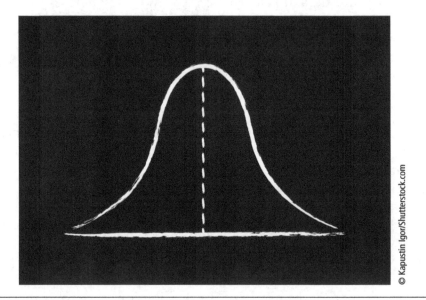

© Kapustin Igor/Shutterstock.com

FIGURE 3.5 The Normal Distribution

Skewed Distributions

Most of the time when we measure things in the world they will form a normal curve, but not all of the time. A "skewed distribution" is not symmetrical. Instead, skewed distributions have scores that group toward one end of a distribution. In a **positively skewed** distribution, there are more low scores than high scores, and so the peak of the curve is toward the lower end of the *x*-axis with the pronounced tail pointing toward the higher values. Figure 3.6 shows what a positively skewed distribution looks like. As you can see, the pronounced tail points away from zero. In a **negatively skewed** distribution, the scores are grouped toward the higher values on the *x*-axis, with the pronounced tail pointing toward zero, as seen in Figure 3.7.

Once you understand statistics on a conceptual level, the skewed distributions will make more sense. For example, look at the negative skew (Figure 3.7). Let us say that this is the distribution of quiz scores in a Statistics class. How did most of the class do? You can answer this question based on the shape of the curve. If you guessed that most of the

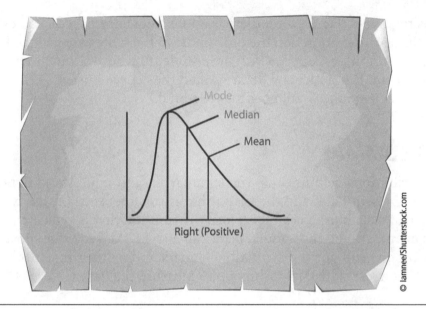

FIGURE 3.6 Positively Skewed Distribution

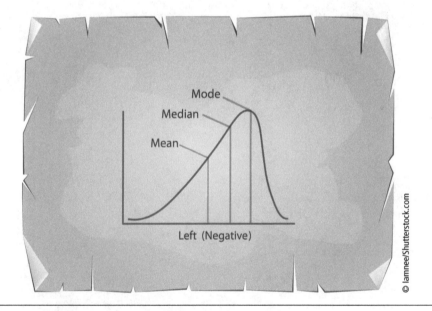

FIGURE 3.7 Negatively Skewed Distribution

class did well you are correct. This is because most of the class (as indicated by the peak in the curve) grades are toward the positive end of the axis. There were some students who did poorly, and it is those low-value, low-frequency scores that pull the pronounced tail toward zero. The next time someone says a test was skewed, positively or negatively, you will know how the class did without seeing any scores or even the distribution.

With a positive skew (Figure 3.6), there is a low frequency of high scores. This means that most of the class did poorly on the quiz (most of the scores are grouped on the lower end of the distribution), except for very few students who did much better than their classmates (these are the students who are called the "curve busters"). This is a positive skew because the pronounced tail goes away from 0.

CHAPTER SUMMARY

Frequency distributions help us organize, summarize, and understand data. Graphs are another way to present data, and the graph that is used depends on the type of data we have. The normal distribution is a theoretical distribution that is symmetrical, with most of the scores in and around the middle. Data can also be skewed, with most of the scores grouped toward one end of the distribution. Understanding the shape of the data aids us in choosing the correct statistical technique to measure the data.

FORMULAS FOR CHAPTER 3

Proportion: $p = \dfrac{f}{N}$

Percentage: $\% = \dfrac{f}{N} \times 100$

1. What are the three things wrong with the following frequency distribution?

Score	f	cf
10	2	0
11	1	3
12	1	4
13	3	7
15	0	7
16	2	9
17	1	10

$$N = 8$$

2. Using the following data (ages in a sample), create a simple frequency distribution (round to two decimal places) and answer the following questions.
   a. How many people are 31 years of age?
   b. How many people are 33 or younger?
   c. What proportion of the sample are 35 years of age?
   d. What percentage of the people in the sample are 36 years of age?
   e. What age is at the 46th percentile?

30	33	32	35	33	33
31	36	31	33	33	31
31	32	33	32	32	33
33	33	35	35	31	30

Score	f	cf	p	cp	%	c%

3. Using the following data (scores on a Statistics test), make a grouped frequency distribution (round to two places) using intervals of 10 and answer the following questions:
   a. How many students scored between 45 and 54 on the test?
   b. What proportion of the students scored between 65 and 74 and below?
   c. What percentage of the students scored between 55 and 64?
   d. In what range of scores did the greatest number of students score?

81	80	80	78	77	76	75	80	80	69	45	60	51
52	55	55	66	54	90	48	50	62	67	78	88	85
50	50	56	77	71	73	75	75	89	67	68	68	88
50	60	67	65	66	55	55	82	84	81	81	84	82

Scores	f	cf	p	cp	%	$c\%$

4. What type of graph would you create when displaying the frequencies of the following:
   a. The different models of cars in a parking lot?
   b. The different heights in a sample of 500 students?
   c. The ranks of professors on a college campus (adjunct, assistant, associate, and full)?
   d. Bodyweights of a sample of 25 students?

MEASURES OF CENTRAL TENDENCY

INTRODUCTION

The frequency distributions and graphs that we explored in Chapter 3 help us to understand the shape of the data. In this chapter, we will explore a more efficient way of understanding the data which is a number (a *statistic*) that will reduce all of the values in a sample or population down to one number, which we will then use to draw conclusions about the set of the values.

For example, imagine we had the bodyweights of 1,000 students on a college campus. That is a lot of data, and looking at those numbers in their raw form would not tell us anything about the typical bodyweight of those students. So what we do is transform (using statistical techniques) all of those bodyweights into one number that best represents all of them.

Here is another example. When I ask one of my students "How are your grades?" they do not usually respond with a qualitative answer (e.g., good, OK, bad). Instead, they give me a number, which is their grade point average. They could answer by telling me every grade they have ever earned, which would be one way to answer the question, but a very inefficient way. So they give me one number that best represents every grade they ever earned.

A **measure of central tendency** is a single number that describes the most typical score in a distribution, or where the center of the distribution tends to be. In Figure 4.1, there are bodyweights for two men's rugby teams (Team A and Team B). Looking at the histograms (this is continuous data), we can see that the most typical bodyweight of team A is 190 lb and that of Team B is 205 lb.

In this chapter, we will explore three different measures of central tendency. The measure of central tendency we use will depend on the type of data, namely, the level of measurement (nominal, ordinal, interval, or ratio) and shape (normally distributed or skewed).

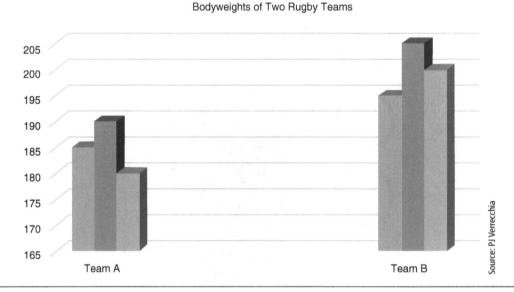

FIGURE 4.1 Bodyweights of Two Rugby Teams

THE MODE

Mode

The mode is the most frequently occurring score in a set of data. For example, let us revisit the examination scores we saw in Chapter 3.

| 90 | 95 | 80 | 80 | 81 | 82 | 92 | 93 | 93 | 95 | 96 | 85 | 85 |
| 85 | 90 | 90 | 88 | 88 | 90 | 90 | 96 | 95 | 84 | 84 | 90 | 90 |

Since the mode is the most frequently occurring score, we count the number of times each score occurs to determine which one is the most frequent. In this case, the mode is 90 (seven students scored 90 on the exam).

The mode can be used with any level of measurement, but we typically use the mode only when we have nominal-level data. In Chapter 3, we saw the favorite color of a sample of 35 people:

Red, blue, blue, brown, black, red, black, green, pink, pink, red, green, blue, orange, blue, red, black, blue, green, red, yellow, red, blue, green, pink, yellow, purple, red, green, black, brown, purple, pink, green, and red.

The mode for this data set would be red because more people said that red was their favorite color (8) than any other color. This brings up an important point about the mode: The mode is not a frequency (the mode here is not 8), but a value that occurs most often.

One advantage to using the mode as the measure of central tendency is that there are no calculations involved; it can be obtained quite easily. Another advantage is its common-sense interpretation. However, there are disadvantages to using the mode. One disadvantage is that the mode ignores data. We learn from the mode that the most frequent test score was 90 and the favorite color was red, but we know nothing of the other test scores and favorite colors. Another disadvantage of the mode is that there can be more than one mode. In Figure 4.2 we see a distribution for grades on a test that has two most frequently occurring scores (this is called a bimodal distribution), and in Figure 4.3 we see a distribution of grades with three most frequently occurring scores (a trimodal

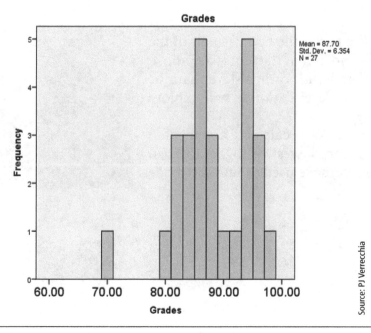

FIGURE 4.2 Grades

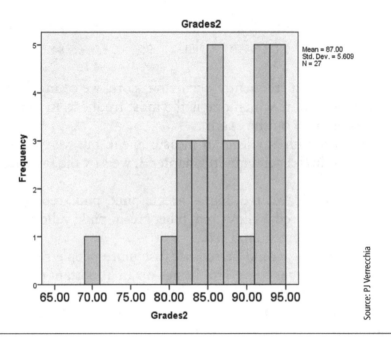

FIGURE 4.3 Grades 2

distribution). Since a measure of central tendency is supposed to be one number that best represents the data, the mode is not always valid. A third limitation is it cannot be used in more sophisticated statistical analysis.

THE MEDIAN

The median is the midpoint of a distribution. It is the score at the 50th percentile because half of the scores fall below it and half of the scores are above it. The median is used with ordinal-level data or skewed interval-/ratio-level data (we will explore why that is shortly).

The first thing you do to calculate the median is rank the values from lowest to highest. When ranked, the exam scores we used to find the mode look like this:

80	84	85	90	90	93	96
80	84	88	90	90	95	96
81	85	88	90	92	95	
82	85	90	90	93	95	

We find the median by using the median locator formula:

$$\frac{n+1}{2}$$

In this case, the sample size (*n*) is 26, so 26 + 1 = 27, divided by 2 equals 13.5. Now the median is not 13.5. What we have to do is count down from the first number to the 13th number and take the average of the 13th and 14th numbers. They are both 90, so the median is 90. For this data set, we would say that a score of 90 is at the 50th percentile; half of the scores fall below 90 and half are above 90.

What if the 13th and 14th scores were two different numbers, say 89 and 90? In that case, we would take the average of 89 and 90 (add them together and divide by 2), and the median would be 89.5. This brings up an important point about the median—it may not be a number in the data set.

One advantage of using the median is that it will always be a single number (unlike the mode). Another advantage is that it is not effected by outliers (extreme values in the data). Disadvantages of the median are that (1) it does not consider all of the scores in a data set and (2) it has limited use in more complex calculations.

THE MEAN

The mean may be the most familiar of the measures of central tendency because it is also known as the average. It is the mathematical center of a set of scores, and we use it when we have normally distributed interval/ratio-level data. We compute the mean just like we compute an average—it is the sum (Σ) of a set of values divided by the number of values involved. The formula is:

© ArpitaJain/Shutterstock.com

$$\bar{X} = \frac{\Sigma X}{n} \qquad \mu = \frac{\Sigma X}{N}$$

$\bar{X}$ is the symbol for the sample mean and the Greek letter μ (pronounced *mew*) is the symbol for the population mean. We differentiate between samples and populations by using the English alphabet with samples and the Greek alphabet with populations.

We know the mode and median for the exam scores from Chapter 3, so let us now find the mean. Take all of the values and add them up, which is 2,307. Then divide that total by the number of scores (26) and the mean is 88.73.

80	84	85	90	90	93	96
80	84	88	90	90	95	96
81	85	88	90	92	95	
82	85	90	90	93	95	

$\Sigma X = 2,307$

$n = 26$

$\bar{X} = 88.73$

The first advantage of the mean is that we are familiar with it (think averages). Another advantage of the mean is that unlike the mode and median, it does not ignore data—it takes every numerical observation in a set of data into consideration. However, a disadvantage to the mean as a measure of central tendency is that it is affected by outliers. Let us consider the same set of data we have been using, but now add an extreme value (outlier) of 200, so the data look like this:

80	84	85	90	90	93	96
80	84	88	90	90	95	96
81	85	88	90	92	95	200
82	85	90	90	93	95	

The mode for this new data has not changed—90 is still the most frequently occurring score. To calculate the median, we change n to 27, add one (28) and divide by two, so the median locator is now 14. However if you count from the lowest value, you will find that the 14th score is still 90, so the median is unchanged as well.

The mean is another story. Adding a value of 200 makes our new total (ΣX) 2,507. Our new n is 27, so $2,507/27 = 92.85$. The mean jumped by 4 with just the addition of one outlier. Since the mean fluctuates if there are outliers in the data, we do not use it with skewed interval- or ratio-level data (we would use the median).

COMPARING THE THREE MEASURES OF CENTRAL TENDENCY

In a normal distribution, all three measures of central tendency will be the same number. Figure 4.4 shows a normal distribution of scores. The score of 75 is the mathematical center (mean), the score at the 50th percentile (median), and the most frequently occurring score (mode). This means that we can conceptualize the shape of a distribution based solely on the values for the three measures of central tendency.

This will not be the case in a skewed distribution. In a negative skew, the mean will have the lowest value, the mode will be the highest, and the median will be in between the two (at the 50th percentile). This means that low frequency, low scores (outliers) are pulling the distribution out of normal, toward 0 (as shown in Figure 4.5). The opposite happens with a positive skew, where the mean will be the highest value, the mode the lowest, and the median in between (as shown in Figure 4.6). In either case with a skewed distribution, the mean gets pulled toward the pronounced tail, while the median will typically fall between the mean and the mode.

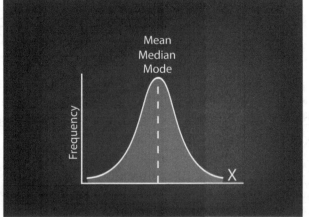

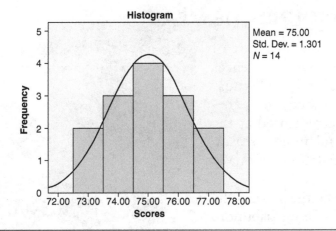

FIGURE 4.4 Histogram

Source: PJ Verrecchia

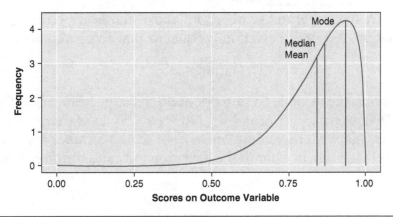

FIGURE 4.5 Negative Skew

Source: From *Data Sense: An Introduction to Statistics for the Behavioral Sciences* by Barton Poulson. Copyright © 2014 by Kendall Hunt Publishing Company. Reprinted by permission.

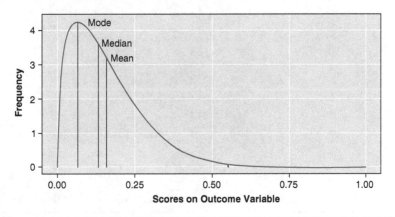

FIGURE 4.6 Positive Skew

Source: From *Data Sense: An Introduction to Statistics for the Behavioral Sciences* by Barton Poulson. Copyright © 2014 by Kendall Hunt Publishing Company. Reprinted by permission.

DEVIATIONS AROUND THE MEAN

The mean is the most used measure of central tendency in Behavioral Sciences research. One reason for this is because many of the things we study in the Behavioral Sciences are measured at the interval/ratio level, and another reason is that we assume that data are normally distributed.

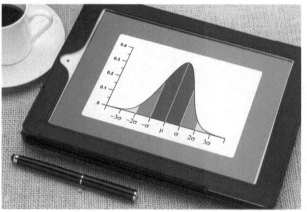

As mentioned earlier, the mean is the mathematical center of a distribution. This means that in a normal distribution there are an equal number of scores above the mean as there are below the mean. The distance that separates scores (both above and below) from the mean is called the score's **deviation.** A score's deviation is equal to that score minus the mean:

$$\text{Deviation} = X - \bar{X}$$

If in a distribution the mean is 65, a score of 60 deviates from it by –5 (60 – 65 = –5). A score of 70 deviates from the mean by 5 (70 – 65 = 5). Deviations can be positive and negative numbers, and what they communicate is a score's *distance* from the mean (which is always positive). The sign in front of the deviation indicates the *direction* from the mean (above or below).

TABLE 4.1 Deviations from the Mean

Score	Mean	Deviation
8	11	−3
8	11	−3
9	11	−2
10	11	−1
11	11	0
11	11	0
12	11	+1
13	11	+2
14	11	+3
14	11	+3
	Sum =	0

This brings us to a third reason why the mean is used so frequently in Behavioral Sciences research. In a normal distribution, the sum of the deviations around the mean always equals zero. For example, Table 4.1 shows the deviations for 10 scores in a normal distribution with a mean of 11.

This is significant because one of the goals of research is to make predictions. If we have to make a prediction about a situation and we have no other information besides the mean, it makes sense to predict the mean. Here is an example.

Let us say I have a student on the basketball team and I am asked to predict how many points she will score in her next five games. If I know that her average points per game is 15, then I will predict that in every game for the next five games she will score 15 points per game. She plays the next five games and scores the following number of points per game: 10, 14, 16, 16, and 18.

At first glance, it looks like I am pretty bad at this since I did not correctly predict the number of points she would score in any game—I was 0 for 5. But over the course of all five games, my errors (deviations from the points that she scored) cancelled themselves out:

Game	Points Scored	Mean Points/Game (my guess)	Deviation
1	10	15	−5
2	14	15	−1
3	16	15	+1
4	16	15	+1
5	18	15	+3
		Sum =	0

Understanding deviations helps further our understanding of statistics conceptually. For example, let us add deviations to Figure 4.4:

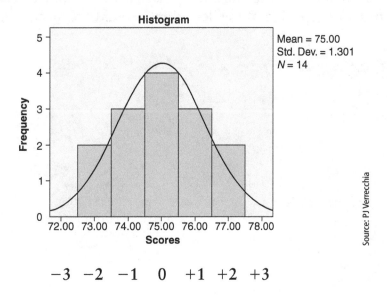

Source: PJ Verrecchia

−3 −2 −1 0 +1 +2 +3

A score of 72 has the lowest value but the highest deviation since it is further away from the mean (75) than all of the other scores except for 78, which also has a deviation of 3. Notice as well that even though the score of 75 has the lowest deviation score, it has the highest frequency since it is in the middle of the normal distribution. This aids in our conceptual understanding of statistics because we could take away the picture of the normal curve, leaving only the deviation scores, and we would still understand the data.

For example, let us say we have a distribution with the following deviation scores:

$$-9, -2, 0, +3, +6$$

Which deviation scored represents the score with the lowest value? If you said –9, you are correct because the negative sign indicates that the raw score is below the mean. Since 9 is the highest number, it means that it is the furthest score from the mean, and the negative sign indicates that it is below (to the left of) the mean. Which score has the highest value? If you said +6, you are correct. The positive sign indicates that it is above the mean, and 6 is the highest of the values above the mean so it is the furthest to the right of the mean.

Here is another one—which deviation score represents the raw score with the highest frequency? If you said 0 you are on a roll, since the score with the highest frequency will be the one closest to the middle. The score with the lowest frequency is –9, since that is the score furthest from the middle.

CHAPTER SUMMARY

A measure of central tendency describes where the center of a distribution tends to be. The measure of central tendency that we use to represent the data depends on the level of measurement (nominal, ordinal, interval, or ratio) and shape (normally distributed or skewed) of the data. A scores' deviation from the mean (the mathematical center of a distribution) is that score minus the mean, and deviations can be above or below the mean. In a normal distribution, the sum of the deviations around the mean will be 0.

FORMULAS FOR CHAPTER 4

Median locator $= \dfrac{n+1}{2}$

$\bar{X} = \dfrac{\sum X}{n}$

Deviation from the mean $= X - \bar{X}$

1. For the following data sets, calculate the mean, median, and mode, and then determine the shape of the distribution:

a.	25	27	29	28	30	27	26	28	31	28	29	
b.	85	85	147	90	86	87	86	91	86	145	92	86
c.	7	59	70	63	61	60	59	65	63	61		
d.	102	91	91	90	90	54	56	70	91			
e.	78	63	98	42	78	64	78	87	72	52		

2. Which measure of central tendency should you compute for the following data sets?
    a. Scores on a 10-point quiz: 8, 8, 9, 7, 8, 10, 9
    b. Favorite musical groups in a sample of 100 students.
    c. Grades: A, A, A–, B, C, A+, D, F, F, C, C–
    d. Scores on a 100-point exam: 85, 86, 87, 88, 88, 85, 86, 38, 89, 88, 85
    e. Jersey numbers: 22, 55, 66, 19, 02, 55, 22, 04, 11, 22

3. On a normal distribution of scores, six participants obtained the following deviation scores: –5, –7, 0, +3, +11 and +1.
    a. Which person obtained the lowest raw score?
    b. Which person's raw score had the lowest frequency?
    c. Which person's raw score had the highest frequency?
    d. Which person obtained the highest raw score?
    e. Rank order the deviation scores in terms of their frequency, lowest to highest.

4. On a normal distribution of scores, five participants obtained the following deviation scores: +1, –3, –2, +7, and +4.
    a. Which person obtained the lowest raw score?
    b. Which person's raw score had the lowest frequency?
    c. Which person's raw score had the highest frequency?
    d. Which person obtained the highest raw score?
    e. Rank order the deviation scores in terms of their frequency, highest to lowest.

CHAPTER 5
MEASURES OF VARIABILITY

© New Africa/Shutterstock.com

INTRODUCTION

In Chapter 4 we discussed calculating a single number that describes an entire distribution of data. That number, the measure of central tendency, is the best representation of all of the scores in a sample or population. A measure of central tendency is a wonderful way to simplify the data, and it helps us understand the distribution without examining every score.

However, a measure of central tendency, as useful as it is, provides an incomplete understanding of the data because we do not know about the differences between the scores. Luckily, there is another statistic that answers that for us—a measure of variability. Variability indicates the extent to which observations differ in value. If there are large differences between scores, the measure of variability will be a higher number, and if there are small differences between scores the measure of variability will be a lower number.

In statistics when we say that scores are **homogeneous**, it means that they are close to each other, and if we say that scores are **heterogeneous** it means that they are apart from each other. These two terms are relative because we use them to compare groups to each other to determine which groups are more alike (homogeneous) or more different (heterogeneous).

In Table 5.1 there are two distributions. Distribution A is an example of perfect homogeneity because each score is exactly the same. The scores in distribution B are also homogeneous because they are close to each other, but distribution A is more homogeneous.

TABLE 5.1 Comparisons of Two Distributions

Distribution A:	5	5	5	5	5	5	5	5	5	5
Distribution B:	5	5	5	6	6	6	6	7	7	7

Table 5.2 displays two distributions that are heterogeneous since the scores are different from each other. Distribution C, however, is less heterogeneous because there is less difference than the scores in distribution D.

TABLE 5.2 Comparison of Two Distributions

Distribution C:	5	10	13	14	17	19	22	28	30	31
Distribution D:	5	22	38	55	66	70	94	114	200	341

Measures of variability communicate three important aspects of the data:

- Measures of variability communicate the *consistency* of the data. Variability and consistency are antonyms, so the greater the variability the less the consistency.
- Measures of variability communicate how *spread out* the scores are from each other. The greater the variability, the greater the spread.
- Measures of variability inform us as to *how well the measure of central tendency describes the data*. For example, the mean does a poorer job of describing a distribution if there is great variability between scores, but a better job if the scores are more consistent.

Here is an example of three sets of data with the same mean but with different variability. If I said that there are three data sets with means of 50, you might think they were all the same numbers, or at least very similar. However, as you can see from Table 5.3, this is not the case.

TABLE 5.3 Comparison of Means

Group A:	25	50	75
Group B:	40	50	60
Group C:	49	50	51

The mean of 50 provides a very good description of the data in group C, a pretty good description of the data in group B, but a poor description of the data in group A. A measure of variability, in addition to the measure of central tendency, provides a more complete description of the data.

The measure of variability we use to describe the differences between scores depends on the type of data we have. There is one measure for nominal- and ordinal-level data, and two measures for interval/ratio-level data. The first measure of variability we will discuss is used with nominal- and ordinal-level data.

THE RANGE

© VarnakovR/Shutterstock.com

The range is the simplest measure of variability, and it is a measure of distance between the highest score and the lowest score in a distribution. The larger the range, the greater the variability. If one distribution has a range of 15 and another has a range of 5, then the second distribution has more consistency (less variability) than the first distribution. The formula for the range is:

$$R = H - L$$

We can apply the range to nominal-level data. Let us say there are two career fairs held on two different college campuses. One career fair had a range of 6 majors and the other had a range of 12. From this, we could say that there was greater variability in majors represented at the second career fair than the first.

To apply the range to ordinal-level data, imagine there are 100 runners in two different races. If the first race had a range of 80, then someone came in 81st and someone came in first (there were some ties). In the second race the range was 50, where someone came in 51st and someone came in 1st (again, there were a number of ties). If you can envision these two races, you can see that the runners in the first race were more spread out (as indicated by the greater range) than the runners in the second race.

While the range is simple to calculate and understand, it is not a very good measure of variability because it only takes into account the most extreme scores while ignoring every other value in a distribution of data. Look at the data in Table 5.4.

TABLE 5.4 Comparison of Two Distributions

Distribution A:	5	5	6	6	6	6	7	7	8	55
Distribution B:	4	10	13	22	34	45	48	49	50	54

The range for each distribution is 50 (55–5 for A and 54–4 for B), but this is misleading because in Distribution A most of the scores are between 5 and 8, but the one extreme score of 55 greatly impacts the range. If there were no numbers and all you knew was that there was a range of 50, you would likely envision a set of scores that were more spread out than they really are. For this reason, the range is not very helpful.

THE VARIANCE

Instead of using only the extreme scores to determine variability (like the range does), a better method would be to compare the distance of the scores from each other and from the center of the distribution. This would inform us of the total variability, or spread, in the data. Since most Behavioral Sciences research involves normally distributed interval/ratio data, the most used measure of central tendency is the mean. There are two measures of variability that are used with the mean, and one is the variance.

© OpturaDesign/Shutterstock.com

We use the variance (and later, the standard deviation) to describe how different scores are from each other and from the mean in a normal distribution of data. As you will recall from Chapter 4, a score's deviation is how far it is from the mean. When we have multiple scores in a distribution, we can understand the spread of the data by calculating the average deviation. The formula for the deviation is

$$\text{Deviation} = X - \overline{X}$$

and the formula for the mean is

$$\overline{X} = \frac{\sum X}{N}$$

so the formula for the average deviation should be

$$\frac{\Sigma\left(X - \overline{X}\right)}{N}$$

where you add all of the deviations in a distribution (raw score minus the mean) and divide by the sample size.

However, this leaves us with a problem. If you recall from Chapter 4, the sum of deviations in a normal distribution always equals 0, since the mean is the mathematical center of any normal distribution. This means that the average deviation would always be 0 (0 divided by any *N* will be 0). The way to correct this is to square the deviations which will make all of the deviations a positive number (a negative times a negative is a positive) so we will have a number we can work with (as shown in Table 5.5). This makes the definitional formula for the variance

$$V = \frac{\Sigma\left(X - \overline{X}\right)^2}{N}$$

Let us take the following set of number of prior offenses to determine the variance.

TABLE 5.5 Squared Deviations

Value	Mean	$X - \overline{X}$	$\left(X - \overline{X}\right)^2$
3	5	−2	4
4	5	−1	1
5	5	0	0
6	5	1	1
7	5	2	4

The sum of the squared deviations is 10 and the sample size is 5, so the variance for this data set is 2. We would say that the average number of prior offenses is 5, with a variance of 2 prior offenses.

But it is not 2 prior offenses. The variance is an inflated number because we had to square the deviations in order to prevent the average deviation from being 0. We would have to say the average number of prior offenses is 5 with a variance of 2 squared prior offenses. Since there is no such thing as a squared offense, we have another problem, the variance does not translate well. The variance is a measure of variability we will see later on in the book, but the conundrum of a squared prior offense brings us to the next measure of variability, the standard deviation.

THE STANDARD DEVIATION

Since the variance is an inflated number due to the squaring of each deviation, in order to make the variability a number that is realistic we simply take its square root. This gives us the true average deviation (or, standard deviation), and the definitional formula is

$$S = \sqrt{\frac{\Sigma(X - \overline{X})^2}{N}}$$

The square root of the variance in the previous problem (2) is 1.41, so we would say that there were an average of 5 prior offenses with a standard deviation of 1.41 prior offenses. This is the average of the deviations in a set of data. While some scores will deviate by more or less (as we saw earlier), 1.41 represents the typical deviation of the scores from the mean and from each other. This also allows us to compare different distributions in terms of their consistency. Remember, higher numbers means less consistency; lower numbers indicate greater consistency. If there are two distributions with a mean of 75, but one has a standard deviation of 4 and the other has a standard deviation of 7, it is safe to conclude that the distribution with the larger deviation has a wider dispersion of scores, has greater variability, and has less consistency in its values.

OTHER FORMULAS

The definitional formula helps us to understand what the variance and standard deviation mean in terms of the average deviation. There are other formulas called computational formulas that allow us to calculate the variance and standard deviation more easily than the definitional formulas. The computational formulas are

$$V = \frac{\Sigma X^2 - \dfrac{(\Sigma X)^2}{N}}{N} \qquad\qquad S = \sqrt{\frac{\Sigma X^2 - \dfrac{(\Sigma X)^2}{N}}{N}}$$

These formulas may look more complicated than the definitional formulas but are actually simpler to use since they require fewer steps. If you recall, to solve the definitional formula we needed four columns of data: the raw sore, the mean, the deviations, and the squared deviations. We only need two columns of data for the computational formulas. Let us use the previous example where we calculated the number of prior offenses. The X column represents the number of prior offenses, and in the X^2 column we square each value.

X	X²
3	9
4	16
5	25
6	36
7	49
Sum = 25	135

Now let us solve for the standard deviation using the computational formula. First, we use the sum of the raw scores squared (ΣX^2) which equals 135. Next, we take the sum of the raw scores ($\Sigma X = 25$) and square that value, which is then divided by the sample size ($N = 5$), and the whole numerator is also divided by the sample size (5). Inserting the numbers, the formula looks like this:

$$S = \sqrt{\frac{\Sigma X^2 - \frac{(\Sigma X)^2}{N}}{N}}$$

Then we solve for S:

$$\text{Step 1: } S = \sqrt{\frac{135 - \frac{(25)^2}{5}}{5}}$$

$$\text{Step 2: } S = \sqrt{\frac{135 - \frac{625}{5}}{5}}$$

$$\text{Step 3: } S = \sqrt{\frac{135 - 125}{5}}$$

$$\text{Step 4: } S = \sqrt{\frac{10}{5}}$$

$$\text{Step 5: } S = \sqrt{2}$$

Answer: $S = 1.41$

The variance is still 2 and the standard deviation is still 1.41.

You will never have a negative variance or standard deviation since there is no such thing as negative variability. There will be no variability at all (0, perfect homogeneity), or there will be some variability (a number greater than 0). This means if you get to step 3 and the number on the right (in this case 125) is larger than the number on the left (in this case 135), then you did something wrong and you should go back and check your work.

When we are determining the variability in a population of scores, we make a slight change to the variance and standard deviation formulas. The formulas themselves do not change but we have to use different symbols because, as you might recall from earlier, when we are working with a sample we use the English alphabet and we use the Greek alphabet when we are working with a population. Therefore, the formulas for the population variance and standard deviation are

$$\sigma^2 = \frac{\sum X^2 - \frac{(\sum X)^2}{N}}{N} \qquad\qquad \sigma = \sqrt{\frac{\sum X^2 - \frac{(\sum X)^2}{N}}{N}}$$

with the Greek letter sigma (σ) meaning the population standard deviation, and sigma squared (σ^2) representing the population variance.

ESTIMATING THE POPULATION VARIABILITY

If we have all of the values for a sample or for a population, we can use the above formulas to calculate the variance and standard deviation. But what if we have a sample and we want to estimate from that sample to a population? To do this, we have to make a slight but important change to the formulas.

The reason we have to change the formula is that a population has greater variability than a sample. For example, if I ask 10 people what their favorite color is I could get (at the most) 10 different answers. But what if I were to ask 20 people the same question? I could get (at the most) 20 different answers. The idea is that the larger the group, the greater the variability.

This makes the formula for the sample variance and standard deviation biased formulas because they will underestimate the amount of variability in a population. To fix this problem, we change the denominator of the formula so it looks like this

$$v = \frac{\sum X^2 - \frac{(\sum X)^2}{N}}{N-1} \qquad\qquad s = \sqrt{\frac{\sum X^2 - \frac{(\sum X)^2}{N}}{N-1}}$$

The lower case "v" for variance and "s" for standard deviation signify that we are estimating from a sample to a population, and $N - 1$ in the denominator is called the **degrees of freedom**. This is the number of scores in a sample that are free to vary. Dividing by the degrees of freedom will inflate the overall number, thus capturing greater variability.

In the number of prior offenses example, the standard deviation we calculated was 1.41. But rather than dividing by the sample size (5) let us divide by the degrees of freedom (in this case $5 - 1$, so 4). This makes the number in step four 2.5, the square root of which is 1.58, which is higher than 1.41, indicating greater variability.

VARIABILITY AND MATHEMATICAL CONSTANTS

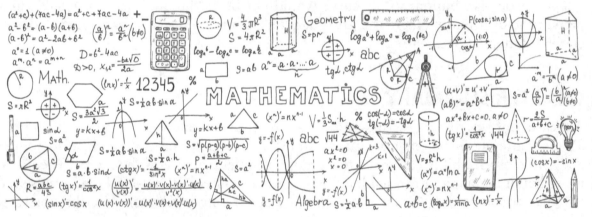

© Nataliia Machula/Shutterstock.com

If we add a mathematical constant to any set of numbers, the size of the numbers will (obviously) get larger, but the variability will remain unchanged. For example, let us calculate the mean and the standard deviation for the following scores:

$$22 \quad 25 \quad 28 \quad 31 \quad 34$$

The sum of all of the values (ΣX) is 140, and the sum of each individual value squared (ΣX^2) is 4010. The sample size is 5, so the mean is 28 and the standard deviation is 4.24.

Now let us add 10 to each score and calculate the mean and standard deviation:

$$32 \quad 35 \quad 38 \quad 41 \quad 44$$

The value for $\Sigma X = 190$, and the $\Sigma X^2 = 7310$. The sample size is still 5, so the mean is 38 but the standard deviation is still 4.24.

The reason for this (this would not be the case if we multiplied by a mathematical constant) is that while the magnitude of the numbers has changed, the distance between the scores has not. Remember, measures of variability measure the distance, or spread, between the scores and the center point. Adding a mathematical constant does nothing to affect the consistency of the scores.

KURTOSIS

The mean of a distribution is the mathematical center around which all of the other scores are located. The standard deviation helps us define "around." Different normal distributions will have different amounts of spread or consistency. **Kurtosis** describes the spread of a normal distribution of data.

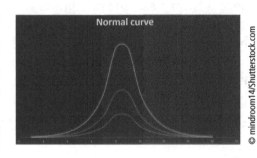

A **leptokurtic** distribution has the lowest amount of variability. A leptokurtic distribution is thin and peaked because the scores are bunched together with very little spread. In a **mesokurtic** distribution, the scores are more spread out than a leptokurtic distribution, so the curve would not be as thin and peaked. Finally, in a **platykurtic** distribution, there is a great deal of spread and the bell shape of the normal curve is flattening out.

CHAPTER SUMMARY

A measure of variability describes the spread, or distance, of the scores in a distribution. The measure of central tendency can be an incomplete descriptor of the data, so adding a measure of variability can make for a more complete picture of the scores. The range is used with nominal- or ordinal-level data, and with interval/ratio-level data we use the variance or standard deviation as our measure of variability. Kurtosis describes the variability of scores in a normal distribution.

FORMULAS FOR CHAPTER 5

$$\text{Range} = H - L$$

$$\text{Sample Variance (computational) } V = \frac{\sum X^2 - \dfrac{(\sum X)^2}{N}}{N}$$

$$\text{Sample Variance (definitional) } V = \frac{\sum(X - \overline{X})^2}{N}$$

$$\text{Sample Standard Deviation (computational) } S = \sqrt{\frac{\sum X^2 - \dfrac{(\sum X)^2}{N}}{N}}$$

$$\text{Sample Standard Deviation (definitional) } S = \sqrt{\frac{\sum(X - \overline{X})^2}{N}}$$

Estimated Population Standard Deviation (computational) $s = \sqrt{\dfrac{\sum X^2 - \dfrac{(\sum X)^2}{N}}{N - 1}}$

Estimated Population Variance (computational) $v = \dfrac{\sum X^2 - \dfrac{(\sum X)^2}{N}}{N - 1}$

Estimated Population Standard Deviation (definitional) $s = \sqrt{\dfrac{\sum (X - \overline{X})^2}{N - 1}}$

Estimated Population Variance (definitional) $v = \dfrac{\sum \left(X - \overline{X}\right)^2}{N - 1}$

1. For the following sets of data, calculate the range, variance, and standard deviation:

   a. 17   24   22   26   28   39   29   28
   b. 3   3   6   6   7   5   5   8   9
   c. 2   1   0   1   8   5   7   8   5   6
   d. 4   4   1   1   0   3   3   6   7   0   9   4

2. Judge Jones handed down the following sentences (in years) to 10 offenders:

   13   9   9   8   12   12   7   6   5   5

   Judge Smith handed down the following sentences (in years) to 10 offenders:

   14   10   9   9   6   10   11   8   6   9

   Calculate the average sentence, standard deviation, and the range for each judge to deter-mine (a) who typically assigns longer sentences and (b) who is the more consistent in their sentencing.

3. You take a random sample of 15 prisoners from the population of offenders incarcerated in two state prisons to find out how many years of education they have completed. Using the data below, calculate the mean and standard deviation to determine (a) which prison has inmates with more years of education, and (b) in which prison are the years of educa-tion more consistent?

Prison A:	11	8	12	9	9	9	10	10
	12	7	9	10	12	11	10	
Prison B:	11	9	11	9	7	9	11	12
	8	8	12	12	8	10	9	

4. Using the data in Table 5.3, which group would you say is
   a. Leptokurtic?
   b. Platykurtic?
   c. Mesokurtic?

CHAPTER 6

Z SCORES AND THE NORMAL DISTRIBUTION

© fatmawati achmad zaenuri/Shutterstock.com

INTRODUCTION

It is assumed in most Behavioral Sciences research that data are normally distributed. Even though the normal curve is a theoretical distribution, we assume that real-world data would approximate a normal distribution. The normal curve has a relationship with the standard deviation. We use the standard deviation to "slice and dice" (more on that later) the normal distribution to produce uniform results that allow for comparisons of different scores and their standing relative to each other. In other words, we can find the proportion or percentage of values that fall between the mean and any other value when we measure the distance in standard deviations.

There are percentages of scores under the normal curve that are revealed to us when we understand the standard deviation. We know that the score at the center of a normal distribution (the mean) cuts the distribution in half, so half (50%) of the scores fall above the mean and half (50%) of the scores fall below the mean. In addition:

- One standard deviation from the mean (on each side) has 34.13% of the scores. That means that in a normal distribution (regardless of kurtosis) $\pm1S$ encompasses 68.26% of all of the scores.

- Two standard deviations from the mean (on each side) have 47.72% of the scores, so in a normal distribution ±2S encompasses 95.44% of all of the scores.
- Three standard deviations from the mean (on each side) have 49.86% of the scores, so in a normal distribution ±3S encompasses 99.72% of all scores.
- Scores beyond ±3S are very infrequent.

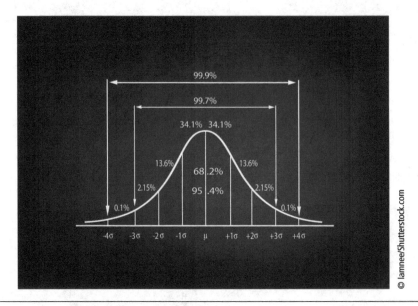

FIGURE 6.1 The Normal Distribution

RELATIVE STANDING

Raw scores in and of themselves do not communicate much without the proper context. We need information to understand them if they are to have any meaning. For example, someone takes an exam and earns a 50. We do not know if this is good, bad, or mediocre unless we know what the total points available were. If they earned 50 out of 50, then they did great. If the total score was 100, then they did poorly. Either way we cannot make much sense out of a raw score in isolation from other information.

However, there is a way to determine how a raw score compares to other scores in a sample or population. What we will do is combine the descriptive procedures from earlier chapters to transform a raw score into something called a z score. Transforming raw scores into z scores allows us to interpret and compare interval- and ratio-level scores.

Three friends, Sue, Bill, and Pat, each take a Statistics exam, and Sue

does better than Bill, who outscores Pat. We can consider this to be ordinal-level data—the students are ranked but we do not know the difference between the rankings. To know that, we need to see their individual exam scores.

On the exam, Sue scored 90, Bill 70, and Pat 40. Now we have interval-level data because we know the differences between the exam scores. Additionally, we know just how much better Sue did than Bill and Pat. What we do not know is how Sue, Bill, and Pat did **relative** to the rest of the students who took that exam. Relative standing is the evaluation of a score relative to the rest of the scores in a sample or population. It lets us know if a score was relatively good, bad, or in-between.

The mean score for this exam was 60. By adding that piece of information we can now compare Sue, Bill, and Pat not only to each other but to the average score for the exam. To do this, we need to determine their exam score's deviation from the mean.

$$\text{Deviation} = X - \overline{X}$$

This is helpful, but we still cannot compare their scores to everyone else in that sample who took the exam. If we convert raw scores into z scores, however, then we can determine each student's standing relative to everyone else who took that exam. This is because a z score tells us how many standard deviations a raw score is above or below the mean. The formula for the z score is

$$z = \frac{X - \overline{X}}{S}$$

where X is the raw score, $\overline{X}$ is the sample mean, and S is the sample standard deviation. The standard deviation is important here because we need to know the variability of the data. Let us say the standard deviation for the exam that Sue, Bill, and Pat took is 10. Now, we can calculate Sue, Bill and Pat's z scores.

$$\text{Sue} = \frac{90 - 60}{10} = 3 \qquad \text{Bill} = \frac{70 - 60}{10} = 1 \qquad \text{Pat} = \frac{40 - 60}{10} = -2$$

Sue's z score is 3 because she scored 3 standard deviations above the mean. Bill's z score is 1 because he scored 1 standard deviation above the mean, and Pat's z score is -2 since she scored 2 standard deviations below the mean. Now we know that not only did Sue outperform Bill and Pat, she outperformed the average score by three standard deviations. Pat did worse than Sue and Bill, but also worse than the average score by two standard deviations.

This brings up an important point. The last chapter stated that you will never have a negative standard deviation. However, you can and will have negative z scores. Any raw score that falls below the mean will have a negative z score even though the raw score is a positive number. The sign in front of the z score (+ or −) tells us if it is above or below the mean, and the number itself tells how far above or below the mean the z score is. (Note: absence of a + sign indicates that it is a positive number.)

We can also use z scores to make comparisons across distributions. Let us say four friends, who are very competitive with each other, take graduate school admissions examinations to get into four different graduate schools: criminal justice, education, business, and literature, and they get the following raw scores.

Criminal Justice: 82
Education: 58
Business: 88
Literature: 73

The business student bragged about earning the best score, but we know that raw scores in and of themselves do not communicate much. So to get a better comparison let us add the mean score for each examination.

Criminal Justice: $\overline{X}$ = 78
Education: $\overline{X}$ = 52
Business: $\overline{X}$ = 100
Literature: $\overline{X}$ = 70

By adding this piece of information, we can see that the business student should not be bragging. Even though their raw score was higher than their three friends, they scored below the average for the business test. The education student's raw score is the lowest, but they outperformed their examination average by more than anyone else.

If we add the standard deviation for each exam, then we can transform each student's raw score into a z score.

Criminal Justice: S = 4
Education: S = 2
Business: S = 6
Literature: S = 1.5

By then applying the z score formula for each raw score, we can compare these scores relative to each other and to every other score in the distribution (Table 6.1).

TABLE 6.1 Comparison of z Scores

	X	$\overline{X}$	S	z
Criminal Justice:	82	78	4	1
Education:	58	52	2	3
Business:	88	100	6	−2
Literature:	73	70	1.5	2

The student with the lowest raw score performed the best on their graduate school admissions exam, relative to their friends. The student taking the literature exam did second best (even though their raw score was second worst), the criminal justice student did the next best, and the business student (with the highest raw score), relatively speaking, did the worst.

THE *Z* DISTRIBUTION

A *z* distribution is produced by transforming all of the raw scores in a distribution into *z* scores. If you look back at Figure 6.1 you will see a *z* distribution. We know that anyone whose raw score was the average had a *z* score of 0. There is a higher frequency of scores closer to the mean, and the further away from the mean we go, higher or lower, the less frequent the scores are because the scores are moving further into the tail regions of the distribution. The *z* distribution has three important qualities:

- The standard deviation of a *z* distribution is always 1.
- The shape of the *z* distribution is the same as the shape of the underlying raw score distribution.
- The mean of a *z* distribution is always 0.

USES OF Z SCORES

The *z*-score is a very powerful statistic because it has a number of uses in Behavioral Sciences research.

Proportions

When we transform raw scores into *z* scores, we can then apply this knowledge to the normal curve to determine the proportion of scores between *z* scores. The student taking the criminal justice examination scored a *z* score of 1, so we know that .3413 (or 34.13%) of the students taking that exam scored between the mean of 78 (a *z* score of 0) and 82 (a *z* score of 1).

We can also determine the proportion of scores between two different nonadjacent values. Let us say someone taking the literature examination earned a raw score of 67 and we want to know the proportion of scores between 67 and 73. This would be the proportion of scores between a z score of –2 and a z score of 2. We know that .9544 of all scores fall between a z of –2 and a z of 2, so we also know that .9544 of the scores fall between a raw score of 67 and a raw score of 73. We determine this by using something called the z tables.

The z tables are found in Appendix A. To find the proportion of an area under the normal distribution, we will use the z tables, a portion of which is represented in Table 6.2.

TABLE 6.2 Partial z Table

A	B	C
+/− Z	Area between Mean and z	Area beyond z In the Tail
1.70	.4554	.0446
1.71	.4564	.0436
1.72	.4573	.0427
1.73	.4582	.0418
1.74	.4591	.0409

If you wanted to find the area that is under the normal distribution for a z score of 1.72, you would first locate the z score of 1.72 in column A ("+/–Z"). Then look in column B, which is the area between the mean (a z score of 0) and the particular z score. You would find the area between the mean and the z score found in column A, which in this case is .4573. This means that .4573 of the scores in a normal curve are between the mean and a z score of 1.72. Since 1.72 is a positive number, this is the area on the right-hand side of the distribution. Column C represents the area beyond a z score into the tail. This contains the proportion of scores under the normal curve that are in the tail regions beyond a z score of 1.72, which in this case is .0427. This means that .0472 of the curve is to the right of 1.72 in the tail.

You will see that there is a positive and negative sign in the z tables, which means that you will decide whether the z score you calculated is a positive or a negative number. If our z score was –1.72, columns B and C would tell us the proportion of scores under the normal distribution on the left-hand side of the curve.

Percentiles

A percentile is the number of scores at or below a certain value, or everything to the left in a normal distribution. We can use the z tables to find a score's percentile, and thus compare their score to everyone else in the distribution.

Let us take the student who took the criminal justice exam. Their raw score of 82 transformed to a z score of 1. We know from looking at the z tables that 34.13% of students scored between a z score of 1 and the mean (a z score of 0).

But since a percentile is everything to the left of a certain score, to determine a percentile for a z score of 1 we have to take that 34.13% between 1 and the mean and add 50% to it (since the mean cuts the distribution in half) to account for all of the scores to the left of the mean. Therefore, someone with a z score of 1 scored at the 84.13 percentile. This means that they did as well as or better than 84.13% of the people in that distribution. Someone whose raw score was at the mean (a z score of 0) scored at the 50th percentile, so they did as well as or better than 50% of the people in that distribution.

To determine the percentile for a z score of 1.74, look at Table 6.2. Since 1.74 is a positive number, it falls to the right of the mean, so the first thing we need to do to determine its percentile is look in column B. A proportion of .4591 of the scores falls in this area under the normal distribution. However, since a percentile is everything to the left of a score we need to add .50 (50%) to account for the entire half of the distribution that is to the left of the mean. The percentile for a z score of 1.74 is 95.41 (.4591 + .50 = .9541, then move the decimal two places to the right).

Let us say we want to know the percentile for a negative z score, like −1.70. In this case a negative number is to the left of the mean, so we do not look in column B. Instead we look in column C. This gives us the proportion of scores to the left (into the tail) of the z score, which is .0446, so a z score of −1.70 is at the 4.46 percentile. This person did as well as or better than 4.46% of the people in that distribution. We do not add .50 to the proportion when the z score is a negative, since it is already to the left of the mean.

Simple Frequencies

Since the z tables inform us of the proportion of scores between two values, we can use that information to determine the number of scores in that area. All we need to know are the z score and the sample size (or population size).

For example, let us say we are interested in knowing not the proportion of scores between 88 and 100 on the business school entrance exam but the number of scores between those two values. We know that 88 transformed to a z score of –2 and 100 is the mean, so the proportion of scores between those values is .4772 (column B in the z tables). To find out how many scores that is, we simply take that proportion (.4772) and multiply it by the sample size (let us say it is 500). Since 500 times .4772 is 238.6, we can say that roughly 238 people scored between 88 and 100 on the business school entrance exam.

If we want to know how many people scored less than 88 on the business school entrance exam, we would again use the z tables. We would find the z score of 2 in column A, and then go to column C to find the proportion of scores to the left (lower than) of that particular z score. The proportion is .0228, so we multiply that value by 500 and get 11.4, which means that the student who scored 88 on the business school entrance examination only did better than about 11 other students.

Raw Score at a Given Percentile

So far we have discussed how to find a z score for a raw score, given the standard deviation and sample mean. We can also reverse this process to find a raw score for a given z score.

Let us say someone taking the criminal justice graduate school entrance examination's raw score transformed to a z score of 3. We know that relatively speaking they did very well. If you find a z score in column A of 3 in the z tables and then go to column B, we see that 49.87% of the exam scores fall between a z score of 3 and the mean. When we add 50% for everything lower than (to the left of) the mean, we know that a z score of 3 is at the 99.87 percentile, which means that this student outscored 99.87% of everyone taking that exam.

We do not know what raw score this is, but we can find out by applying the following formula

$$X = (z)(S) + \overline{X}$$

where X is the unknown raw score, z is the given z-score, S is the standard deviation, and $\overline{X}$ is the sample mean. If we input the mean and standard deviation for the criminal justice examination (refer to Table 6.1) and a z score of 3, we find that the corresponding raw score is 90:

$$X = (3)(4) + 78$$
$$X = 12 + 78$$
$$X = 90$$

CHAPTER SUMMARY

The z score is a very powerful statistic. We use z scores to understand a raw score's relative standing, that is, how it compares to other scores in a distribution of data. It communicates more about the data than a raw score, and z scores can be used to determine proportions, percentiles, simple frequencies, and raw scores at a given percentile.

FORMULAS FOR CHAPTER 6

$$z = \frac{X - \overline{X}}{S} \qquad\qquad X = (z)(S) + \overline{X}$$

1. John earned 85 ($\overline{X}$ = 90, S = 5) on his Statistics exam. His friend Sarah earned a 65 ($\overline{X}$ = 55, S = 5) on her French exam. Should John be bragging about how much better he did? Why or why not?

2. For the following data, compute the mean and standard deviation:

   5   4   6   7   8   8   7   6   10   10   9   3

   a. Compute a z score for a raw score of 6.
   b. Compute a z score for a raw score of 5.
   c. Compute a z score for a raw score of 9.
   d. Compute a z score for a raw score of 10.

3. For the data in question 2, find the raw score that corresponds to the following:
   a. $z = +1.22$
   b. $z = -.48$
   c. $z = +2.11$
   d. $z = -.52$

4. Which z score in the following pairs corresponds to the lower raw score?
   a. $z = +1.1$ or $z = +2.0$
   b. $z = -2.5$ or $z = -1.9$
   c. $z = -.55$ or $z = -.25$
   d. $z = 0.0$ or $z = -1.5$
   e. $z = 0$ or $z = +1.2$

5. In a normal distribution, what proportion of all scores would fall into each of the following areas?
   a. Between the mean ($z = 0$) and $z = +1.75$
   b. Below $z = -2.11$
   c. Between $z = -1.10$ and $z = +2.10$
   d. Above $z = +1.84$ and below $z = -2.00$

6. Below you will find a table with the mid-term examination grades for five students, along with the class average and standard deviation. Complete the table by including their z scores as well as the percentile:

Student	Grade (X)	Class $\overline{X}$	Class S	z Score	Percentile
A	85	75	5		
B	80	55	10		
C	70	75	10		
D	65	70	15		
E	40	30	20		

7. Below you will find a table with mid-term examination z scores for five students. Complete the table by calculating their raw scores:

Student	z Score	Class $\overline{X}$	Class S	Raw Score (X)
A	1.6	65	5	
B	3.0	58	3	
C	−.80	73	10	
D	−1.5	80	5	
E	1.1	60	9	

8. In a distribution in which $\overline{X} = 50$, $S = 8$, and $N = 500$
   a. What is the proportion of scores between 40 and the mean?
   b. How many participants scored between 40 and the mean?
   c. What is the percentile of someone who scored 40?

9. Sophie is taking five courses this semester and earned the following scores on her final examinations:

Course	Sam's Score	Class Mean	Class Standard Deviation
Sociology	82	78	4
Criminal Justice	58	52	2
Microeconomics	89	89	5
Ethics	73	70	6
Seminar	75	70	3

   a. What were Sophie's z scores for each final examination?
   b. On which examination did she do the best, relative to her other classes?
   c. On which examination did she do the worst, relative to her other classes?

10. The final exam in Pharmacology had a mean score of 75 and a standard deviation of 10. Below are listed the raw scores for three students. What are their equivalent z scores?

Stacie $X = 86$ $z =$
Stanley $X = 72$ $z =$
Sarah $X = 92$ $z =$

Below are listed the z scores for three students taking the same Pharmacology final. What are their equivalent raw scores?

Patricia $z = +1.00$ $X =$
Sean $z = -.50$ $X =$
Barbara $z = 0$ $X =$

CHAPTER **7**

CORRELATION

© ibreakstock/Shutterstock.com

INTRODUCTION

Correlation requires scores from two variables (correlation means "co-relationship"), and correlational research involves using statistical procedures to analyze whether a systematic pairing of variables exists. A correlation coefficient is a descriptive statistic that summarizes and describes the characteristics of a relationship between the two variables. No matter how many pairs of numbers you are analyzing, a correlation coefficient transforms all of them into a single statistic. We then examine and interpret that statistic to better understand the nature of the relationship. It has to be emphasized that though there may be a relationship between two variables, we cannot say that one causes the other. However, if there is a relationship between the variables, changes in the one will be associated with changes in the other.

CORRELATIONAL RESEARCH

In Chapter 2, we discussed that a relationship between two variables exists when a change in one is associated with a change in the other. When we find a relationship, we want to know more about that relationship—how consistent it is, what the pattern is, and what the direction of the relationship is. The best way to do this is to compute a correlation coefficient that summarizes and describes these important aspects of a relationship. A correlation coefficient simplifies complex relationships.

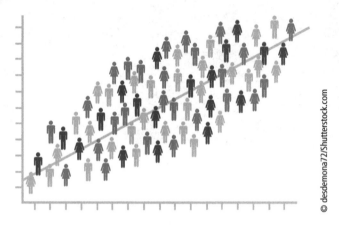

© desdemona72/Shutterstock.com

Let us assume that the more hours a piano player spends practicing the piano, the higher the score they will earn on their final examination in piano class. We could test this in an experiment where we manipulate the number of hours people are allowed to practice. We have some students practice for 1 hour a week, others for 2 hours a week, and so on up to 5 hours per week. Then we would have each participant in this study perform a certain song which would be judged by an expert teacher for a score of 0 to 10, where 10 is the best score and 0 is the worst.

The independent variable (X) in this study is the number of hours practicing the piano, and the dependent variable (Y) is the score earned on the test. In this situation, we are manipulating the independent variable so this would be a true experiment.

However, if you recall from Chapter 2 there is a difference between experiments and correlational studies. In a correlational study you do not manipulate the independent variable. So for this example, we would collect data on the dependent variable, which is the score on the piano test, and then ask each student how many hours they practiced. We would then put the students into different categories (practiced for one hour, practiced for 2 hours, etc…), and then examine the results to see if there is a relationship between time spent practicing and score on the test. The results are presented in Table 7.1.

As you can see from Table 7.1, there is an obvious relationship between hours of practice and score on the piano test: the more one practices, the higher the score they earn on the test. There is a consistent relationship because as the hours of practice (X) increases, so does the score on the test (Y).

An important difference between correlational and experimental research is that in an experiment the X variable is called the independent variable and the Y variable is the dependent variable. In an experiment, we would say that practicing the piano more causes a higher score on the piano test. If we do not conduct a true experiment where the independent variable (hours of practice) is manipulated, it is a correlational study, where

TABLE 7.1 Hours of Practice and Score on Piano Test

Hours of Practice (X)	Score on Piano Test (Y)
1	1
1	1
2	1
2	2
3	5
3	5
4	6
4	8
5	8
5	10

we observe the occurrence of two variables in a natural setting. In correlational research, the X variable is referred to as the **predictor variable** and the Y variable is referred to as the **criterion variable**.

SCATTERPLOTS

In correlational research, we examine relationships using a scatterplot. A scatterplot is a graph that shows the location of data points for a pair of X–Y scores. It is a visual representation of the relationship between two variables. Figure 7.1 is a scatterplot for the data in Table 7.1. It demonstrates that as people practice the piano more, they get higher scores on the piano test.

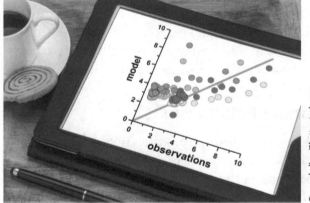

The scatterplot shows that one hour of practice is associated with a score of one on the piano test. We can also see that 2 hours of practice is associated with two different scores on the piano test (2 and 3). Looking at Figure 7.1, we can see the relationship between the predictor and criterion variables, but to understand more about the relationship we need to calculate a correlation coefficient. A correlation coefficient communicates two important aspects of the relationship: the **type** of the relationship and the **strength** of the relationship.

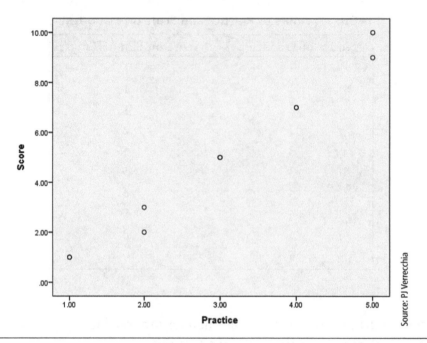

Source: PJ Verrecchia

FIGURE 7.1 Scatterplot for Hours of Practice and Score on Piano Test

TYPES OF RELATIONSHIPS

The type of relationship in a set of data is represented by the direction of the relationship. Generally there are two different types of relationships: **linear** and **curvilinear**.

Linear Relationships

A linear relationship is one in which the data in a scatterplot forms a straight line. This is because as the X scores change, the Y scores also change, but only in one direction. There are two types of linear relationships: positive and negative.

© ESB Professional/Shutterstock.com

A **positive linear relationship** is one in which as the X variable increases, the Y variable increases as well. Take, for example, the data in Figure 7.1. As the X variable increases, the Y variable increases, so higher values for time spent practicing (predictor variable) are associated with higher scores on the piano exam (criterion variable). The data are moving in the same direction, which gives us a positive linear relationship.

A positive linear relationship can also be stated in the negative. What this means is that we could also describe the relationship in Figure 7.1 as follows: the less one practices the piano, the lower the score they earn on the piano test. This also reflects that the data are moving in the same direction: lower values of *X* are associated with lower values of *Y*. In other words, when high values of one variable are associated with high values of another variable, or low values of one variable are associated with low values of another variable, they are positively associated. Examples of positive relationships include amount of rainfall and growth in a garden, height and weight, and time spent exercising and calories burned.

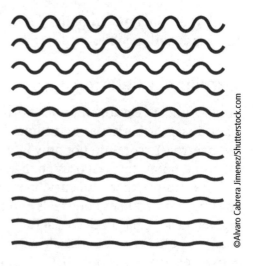

In a **negative linear relationship,** the data move in opposite directions. As the values for the X variable increase, the values for the Y variable decrease. Figure 7.2 shows a scatterplot demonstrating the relationship between the amount of time one spends social networking and their score on the piano test.

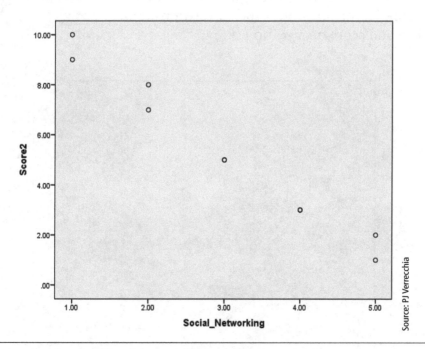

FIGURE 7.2 Scatterplot for Social Networking and Score on Piano Test

Here, we can see that the more time individuals spent social networking, the lower they scored on the piano test. We could also say that the less time one spends on social networking, the higher the score on the piano test. Either way, the data are moving in the opposite direction. When higher values of the one variable are associated with lower values of another variable, and vice versa, there is a negative (or inverse) association. Examples of a negative relationship include increase of speed in a car and time it takes to reach a destination, amount of food eaten and feeling of hunger, and number of absences in a class and grade earned in that class.

Curvilinear Relationships

A curvilinear relationship means that the pattern of the data does not form a straight line. Here, as the X scores change, the Y scores do not only increase or only decrease. At some point, the Y scores change direction.

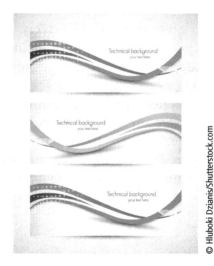

For example, there is a curvilinear relationship between the amount of anxiety one feels and their performance on a given task. Some anxiety is actually beneficial for performance—it focuses us and makes us eliminate distractions. However, anxiety is only beneficial to a point; too much anxiety can be detrimental to our performance (when athletes succumb to too much pressure, we say that they "choked"). Figure 7.3 demonstrates such a relationship between anxiety and score on a piano test.

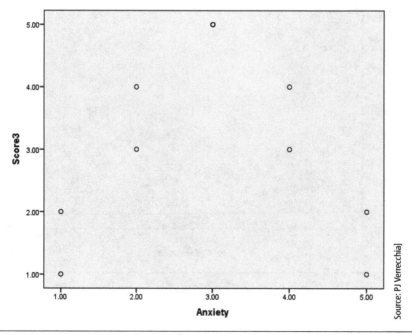

FIGURE 7.3 Scatterplot for Anxiety Level and Score on Piano Test

In Figure 7.3, we can see that as anxiety increases so do the scores on the piano test, but only to a point. Anxiety levels beyond 5 are associated with lower scores on the piano test, which means that as anxiety levels keep increasing, scores on the piano test decrease. The scatterplot shows a curve in the data, hence the name a curvilinear relationship.

Curvilinear relationships can work in the other direction. Consider the relationship between the age of a car and its value. Years ago I was told by a car salesman, just after I bought the car from them, that value of a new car decreases the minute you drive it off the lot. And, generally speaking, as cars get older they go down in value (why he chose to tell me this before the ink was dry on the paperwork remains a mystery to me). However, there are some cars that are called "antique" or "classic" cars that actually see their value increase the older they get. This is illustrated in Figure 7.4. With this type of curvilinear relationship, lower X values (age of the car) are associated with higher Y values (the value of the car). As X increases, Y decreases, but at some point as X keeps increasing, Y starts to increase as well. In other words, the variables are positively or negatively related within a certain range of values, and then they become negatively or positively related in another range of values; they are neither always increasing nor always decreasing.

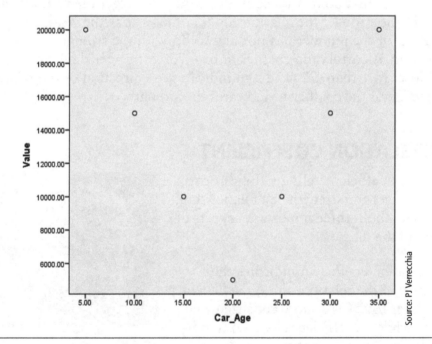

FIGURE 7.4   Scatterplot for Age of a Car and Value

Of course there is another possibility in which there is no relationship at all, where high or low values of one variable are just as likely to be associated with high or low values of another variable throughout the entire range of scores. Here the variables are unrelated or uncorrelated, and examples include eye color and grade point average, and number of candy bars eaten and score on an eye exam.

STRENGTH OF A RELATIONSHIP

The strength of a relationship describes the consistency of a relationship. The size of the correlation coefficient that we compute indicates the strength of the relationship, regardless of its direction (positive or negative). The largest value you can have for a correlation coefficient is 1, which is a perfect relationship. A perfect relationship means that as the X variable changes, the Y variable changes in the exact same fashion. For example, if there was a perfect relationship between hours spent practicing the piano and score on a piano test, then every person who practiced for the same amount of time would get the same exact score on the test.

The correlation coefficient will be any value between -1 and $+1$, and a correlation coefficient of 0 indicates that no relationship is present. The larger the value of the coefficient (the closer to 1), the stronger the relationship. The closer to 0, the weaker the relationship. The sign in front of the number has nothing to do with the strength of the relationship, only the direction. A correlation coefficient of $-.73$ is stronger than a coefficient of $+.22$ since .73 is closer to 1 than .22. The sign indicates the direction of the relationship, and the absolute value of the coefficient indicates the strength.

THE CORRELATION COEFFICIENT

The correlation coefficient tells us the direction (positive or negative) and strength of a relationship, and it informs us about three important aspects of a relationship in the data.

- The correlation coefficient indicates the **consistency** of the relationship. A coefficient that is closer to ± 1 is a more consistent relationship, while a coefficient closer to 0 indicates a less consistent relationship.
- If we know the consistency of a relationship, we also know the **variability** of the relationship.
- The correlation coefficient communicates the **accuracy** of our predictions. A coefficient of ± 1 means that our predictions will be very accurate because knowing the value of an X score means we also know the value (or close to it) of the corresponding Y score. The weaker the relationship, the less accurate our predictive ability.

We will discuss two correlation coefficients. The one we use depends on the level of measurement of the criterion (Y) variable. When the criterion is measured at the interval/

ratio level, we will use the Pearson's Product Moment Correlation Coefficient, more commonly referred to as the Pearson's r. If the criterion is measured at the ordinal level, then we will use the Spearman's rho. We will first turn our attention to the Pearson's r.

Pearson's r

The Pearson's r is probably the most used correlation coefficient in Behavioral Sciences research. It describes a linear relationship between two variables when the Y variable is measured at the interval/ratio level. The Pearson's r measures how consistently each value of Y is for its paired X value. To compute the Pearson's r, we use the following formula (that we have already seen in Chapter 1).

$$r = \frac{N(\Sigma XY) - (\Sigma X)(\Sigma Y)}{\sqrt{\left[N(\Sigma X^2) - (\Sigma X)^2\right]\left[N(\Sigma Y^2) - (\Sigma Y)^2\right]}}$$

While this formula may look daunting, the secret to computing the Pearson's r (or any statistic for that matter) is in organizing the data. To solve the Pearson's r we need five columns of data.

ΣX means that we add up all of the X values, and ΣY means that we add up all of the Y values. For ΣX^2, we need to square each X value and then add those values (just like we did in the computational formula for the standard deviation). For ΣY^2 we will square each Y value and then add those values together. Finally, for ΣXY we multiply each X value times its corresponding Y value, and then add those numbers together. Let us do this using the data in Table 7.1, which examines the relationship between hours of practice (X) and score on a piano test (Y):

X	Y	X_2	Y_2	XY
1	1	1	1	1
1	1	1	1	1
2	1	4	1	2
2	2	4	4	4
3	5	9	25	15
3	5	9	25	15
4	6	16	36	24
4	8	16	64	32
5	8	25	64	40
5	10	25	100	50
$\Sigma X = 30$	$\Sigma Y = 47$	$\Sigma X^2 = 110$	$\Sigma Y^2 = 321$	$\Sigma XY = 184$

N means number of pairs, so now we plug the numbers into the formula:

$$\text{Step 1: } r = \frac{10(184) - (30)(47)}{\sqrt{\left[10(110) - (30)^2\right]\left[10(321) - (47)^2\right]}}$$

$$\text{Step 2: } r = \frac{1,840 - 1,410}{\sqrt{[1,100 - 900][3,210 - 2,209]}}$$

$$\text{Step 3: } r = \frac{430}{\sqrt{[200][1,001]}}$$

$$\text{Step 4: } r = \frac{430}{\sqrt{200,200}}$$

$$\text{Step 5: } r = \frac{430}{447.44}$$

$$r = +.96$$

Our last step is to interpret what this number means. Remember, the closer to ± 1, the stronger the relationship, so a Pearson's r of +.96 means that we have a very strong, positive relationship. We interpret the strength of the relationship using the following criteria (Table 7.2):

TABLE 7.2 Strength of Correlations

Value	Strength
$\pm$.00 –.20	Very weak relationship
$\pm$.21 –.40	Weak relationship
$\pm$.41 –.70	Moderate relationship
$\pm$.71 –.90	Strong relationship
$\pm$.91 –1.00	Very strong relationship

There are three parts to successfully calculating and interpreting a correlation coefficient. First compute the correct number, which in this case is +.96. Second, we interpret what that number means. For this data, it is a very strong, positive relationship. The last step is to always refer back to the variables. This is crucial because while it is unknown whether you will calculate another correlation coefficient, you will see correlation coefficients in your professional career and you will have to know what they mean. For this example, we would say that the more one practices the piano, the higher score they will earn on the piano test.

Spearman's Rho

The second correlation coefficient we will discuss is the Spearman's Rank-Order Correlation Coefficient, also known as the Spearman's rho. Spearman's rho is used when we have pairs of ordinal (ranked)-level data and want to describe a linear relationship between them. Spearman's rho describes the extent that one (ranked) variable is associated with another (ranked) variable. If both scores have the same rank, then the value of Spearman's rho will be

+1. If all of the scores have an opposite rank, then the value will be −1. If there is only a degree of consistency between the rankings, then the value of Spearman's rho will be between +1 and −1. The formula for the Spearman's rho is as follows:

$$r_S = 1 - \frac{6\left(\sum D^2\right)}{N\left(N^2 - 1\right)}$$

With Spearman's rho, we are comparing the differences between rankings (D), and N is the number of pairs. To explain, let us look use the example we used for the Pearson's r, but in a different way.

Instead of the piano players earning a score on a piano test, let us say that they were ranked by two different judges, independent of each other, on their ability to play the piano. Spearman's rho will allow us to gauge the consistency of the judges' rankings. So here, we are not looking at the relationship between hours of practice and score on a piano test, but rather the amount of agreement between the judges in terms of who was the best (a ranking of 1), who was the worst (a ranking of 10), and everything in between. The rankings look like this:

Player	Judge A Ranking	Judge B Ranking
1	2	3
2	1	2
3	4	1
4	3	4
5	5	7
6	6	5
7	8	6
8	10	9
9	7	8
10	9	10

To solve for Spearman's rho, we need to arrange the data to reflect not only the rankings but also the differences (D) between the rankings, which we get by subtracting the second number (in this case judge B's ranking) from the first number (judge A's ranking) (smaller differences mean more consistency). We then square each of the differences (D^2) between the rankings to eliminate any negative numbers, and then we add those squared differences together (ΣD^2).

Judge A Ranking	Judge B Ranking	D	D_2
2	3	−1	1
1	2	−1	1
4	1	3	9
3	4	−1	1
5	7	−2	4
6	5	1	1
8	6	2	4
10	9	1	1
7	8	−1	1
9	10	−1	1
			$\Sigma D^2 = 24$

Then we plug the numbers into the formula:

$$r_S = 1 - \frac{6\left(\Sigma D^2\right)}{N\left(N^2 - 1\right)}$$

Step 1: $r_S = 1 - \dfrac{6(24)}{10\left(10^2 - 1\right)}$

Step 2: $r_S = 1 - \dfrac{144}{10(100 - 1)}$

Step 3: $r_S = 1 - \dfrac{144}{10(99)}$

Step 4: $r_S = 1 - \dfrac{144}{990}$

$$\text{Step 5: } r_S = 1 - .15$$

$$r_s = +.85$$

We interpret Spearman's rho like we do Pearson's r in terms of the direction and magnitude of the calculated number: this is a strong (refer back to Table 7.2), positive relationship. The difference is in what we say next. A Spearman's rho of + .85 means that there was strong agreement between the judges, and the piano players that judge A ranked higher, judge B also ranked higher. You could also say that the piano players that judge A ranked lower, judge B also ranked lower. Remember, in a positive relationship the data move in the same direction.

CHAPTER SUMMARY

Correlation (co-relationship) requires scores from two variables, and we use correlational research to analyze the relationship between those two variables. Linear relationships can be positive, where the variables move in the same direction, or negative, where they move in opposite directions. Relationships can also be curvilinear, where a relationship starts out as positive but at some point becomes negative (and vice versa). In addition to the direction of a relationship, correlation coefficients allow us to determine the strength of a relationship, which aids in our predictive ability. The correlation coefficient we use to analyze relationships depends on the level of measurement of the data. If the criterion variable is measured at the interval/ratio level we use the Pearson's r, and if we are using ordinal-level data we use the Spearman's rho.

FORMULAS FOR CHAPTER 7

Pearson's r:

$$r = \frac{N(\Sigma XY) - (\Sigma X)(\Sigma Y)}{\sqrt{\left[N(\Sigma X^2) - (\Sigma X)^2\right]\left[N(\Sigma Y^2) - (\Sigma Y)^2\right]}}$$

Spearman's rho:

$$r_S = 1 - \frac{6(\Sigma D^2)}{N(N^2 - 1)}$$

1. From a sample of 10 college students, determine the relationship between hours spent studying and score on a statistics quiz by computing and interpreting the appropriate correlation coefficient for this data.

Hours Studying (X)	Score on Quiz (Y)
0	4
1	7
0	3
2	5
3	9
3	8
5	10
0	3
1	3
3	7

2. Below you will find the class rankings for 10 students for their freshman and sophomore years. Calculate and interpret the appropriate correlation coefficient for this data.

Class Rank Freshman	Class Rank Sophomore
1	2
2	1
4	6
5	10
7	8
3	9
6	7
10	3
8	9
9	5

3. A statistics class meets once a week over a 16-week semester. Below you will find the grades of a class of 15 and the number of classes missed during the semester. Describe the relationship between grades and missing classes.

Number of Missed Classes(X)	Grade in Class(Y)
0	4
2	3.5
2	4
3	3.5
3	2
1	2.5
0	3
3	3
3	3
0	4
4	3
0	3.5
5	1
6	0
4	2

4. A researcher wants to determine if there is a relationship between the number of grammatical mistakes a student makes in a term paper and the student's satisfaction with their performance (measured at the ratio level). Using the following data, calculate and interpret the Pearson's *r* to conclude if there is a relationship between number of mistakes and satisfaction.

Mistakes (X)	Satisfaction (Y)
8	2
7	1
3	7
5	4
6	3
9	1
4	6
4	6

5. Using the following ordinal-level data, determine if there is a relationship between a student's class rank and where they rank in terms of number of days they have been late to school (a ranking of 1 means that they have been late more than anyone else, 2 means fewer incidences of lateness, and so).

Class Rank	Ranked Lateness
1	9
2	10
10	4
9	1
8	3
6	6
5	5
7	2
3	7
4	8

6. For each of the following relationships, state whether the correlation would be positive or negative.
   a. Level of education and annual salary.
   b. Amount of rainfall (in inches) and attendance at an outdoor football game.
   c. Number of hours watching movies and grade point average.
   d. Number of hours spent studying and grade point average.
   e. Miles driven and amount of fuel used.

7. Determine the relationship between number of sick days taken in the last month and the number of units produced in a factory.

Number of Sick Days (X)	Number of Units Produced (Y)
2	12
8	4
0	15
3	10
5	8
4	7

8. Determine the relationship between the number of books read in a month and score on a spelling quiz.

Number of Books Read (X)	Score on Spelling Quiz (Y)
0	1
2	3
1	0
3	1
4	4
5	6

9. A track and field coach ranked the discus throwers on her team on their level of preparedness for the next track meet. Determine the relationship between her rankings and the rankings of the throwers at the finish of the meet.

Coach's Ranking of Preparedness (X)	Finish at Meet (Y)
6	6
3	1
4	4
5	5
2	3
1	2

CHAPTER 8

LINEAR REGRESSION

© SNeG17/Shutterstock.com

INTRODUCTION

A relationship between variables exits when scores on one variable (X) are paired with scores on another variable (Y). For example, the first practice problem in the preceding chapter sought to establish if there was a relationship between the number of hours students study and score on a quiz. If there is a relationship, then we could predict quiz scores based on how many hours someone studied. The procedure for making predictions is **linear regression.**

When I applied to schools to earn my master's degree, I had to take an entrance exam called the Miller's Analogies Test (MAT). (This would not have been necessary if my undergraduate grades had been better, so let that be a lesson to you—hit the books.) The reason this university used an entrance exam is because they found a positive correlation between scores on the test and grades—students who earned high scores on the MAT tended to earn better grades. Regression techniques can be used to a predict student's grades based on their MAT score. Employers who have applicants take a preemployment screening test are using the same logic, predicting success at the job based on test scores.

103

THE REGRESSION LINE

Pearson's r is the statistic that describes the direction and strength of a relationship. The **regression line** is a line that summarizes a relationship. A regression line is a line drawn through a scatterplot in a way that is a representation of the data in the scatterplot.

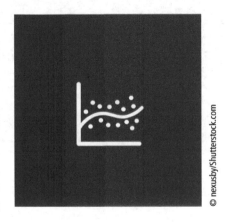

If you look at Figure 8.1 you will see a scatterplot that shows a positive relationship (as X scores are increasing, Y scores are also increasing). Now imagine Figure 8.1 with the dots of the scatterplot erased and only the regression line left. With only the line, we would still know that the data represent a positive linear relationship because the line is a diagonal line moving upward from left to right.

The regression line is also called the **line of best fit** because it is a line that is the best representation (best fits) of the data. It does not have to touch each data point (it is not a polygon), but it does have to do the best job of summarizing the relationship. The regression line helps us understand regression techniques, but to make predictions we have to use the linear regression equation.

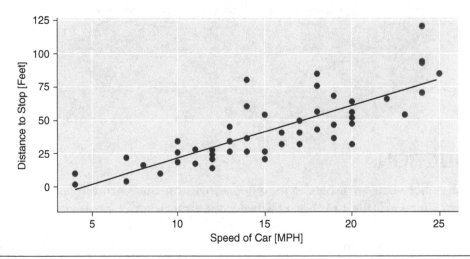

FIGURE 8.1 Scatterplot for Speed of Car and Distance Traveled

Source: From *Data Sense: An Introduction to Statistics for the Behavioral Sciences* by Barton Poulson. Copyright © 2014 by Kendall Hunt Publishing Company. Reprinted by permission.

THE LINEAR REGRESSION EQUATION

We compute the **linear regression equation** to predict a criterion variable (Y' is the symbol for predicted Y) given a known value of a predictor variable (X). The linear regression equation is the equation for a straight line that describes a relationship and produces the value of *Y'* at any *X*. The regression equation involves two characteristics of the regression line—the slope and the *y* intercept.

The **slope** is a statistic that indicates the direction of the relationship. When a relationship is positive, then the slope will look like the regression line in Figure 8.1. As the *X* scores increase, the *Y* scores also increase. If the relationship is negative, then the slope will move downward from left to right (the opposite of Figure 8.1). Finally, when there is no relationship then the slope will be a horizontal line. As the *X* scores increase, the *Y* scores do not change.

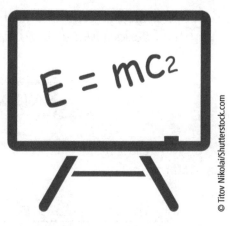

The **y intercept** is the second important characteristic of the linear regression equation. The *y* intercept is the value of *Y* at the point where the regression line crosses (or intercepts) the *Y* axis. The *y* intercept also describes how the *Y* scores change by telling us the starting point at which they start to change.

So, the linear regression equation combines the slope (*b*), *y* intercept (*a*), and a predictor value (*X*) to calculate a predicted *Y* (*Y'*):

$$Y' = b(X) + a$$

As an example, a company that makes desks for colleges uses a preemployment test designed to measure proficiency at desk assembly. The test ranges from a low of 1 (very little proficiency) to 5 (highest level of proficiency). To validate this test, researchers set up a scenario whereby people take the test, and then are tasked with assembling as many desks as they can in 1 hour. The predictor variable (*X*) is score on the test, and the criterion variable (*Y*) is the number of desks assembled in 1 hour. The researchers recruit 10 volunteers and get the following results:

Score on Test (X)	Desks Assembled (Y)
1	2
2	2
5	6
5	7
4	4
4	5
3	2
1	1
1	2
2	3

The first step is determining if there is a relationship between score on the test and number of desks assembled, so we need to compute Pearson's r:

$$r = \frac{N(\Sigma XY) - (\Sigma X)(\Sigma Y)}{\sqrt{\left[N(\Sigma X^2) - (\Sigma X)^2\right]\left[N(\Sigma Y^2) - (\Sigma Y)^2\right]}}$$

As you recall from Chapter 7, the key to computing Pearson's r is how we organize the data:

X	Y	X_2	Y_2	XY
1	2	1	4	2
2	2	4	4	4
5	6	25	36	30
5	7	25	49	35
4	4	16	16	16
4	5	16	25	20
3	2	9	4	6
1	1	1	1	1
1	2	1	4	2
2	3	4	9	6
$\Sigma X = 28$	$\Sigma Y = 34$	$\Sigma X^2 = 102$	$\Sigma Y^2 = 152$	$\Sigma XY = 122$

And then we plug the numbers into the equation:

Step 1: $r = \dfrac{10(122) - (28)(34)}{\sqrt{\left[10(102) - (28)^2\right]\left[10(152) - (34)^2\right]}}$

Step 2: $r = \dfrac{1,220 - 952}{\sqrt{[1020 - 784][1520 - 1156]}}$

Step 3: $r = \dfrac{268}{\sqrt{[236][364]}}$

Step 4: $r = \dfrac{268}{\sqrt{85,904}}$

Step 5: $r = \dfrac{268}{293.09}$

$r = +.91$

This is a very strong, positive relationship. The higher the score on the test, the greater number of desks assembled. Now that the direction and strength of this relationship have been established, and since this is a very strong relationship, we can now predict how many desks will be assembled based on a score on the proficiency test. To make a prediction, we will use the linear regression equation, and we start by calculating the slope:

$$b = \frac{N(\Sigma XY) - (\Sigma X)(\Sigma Y)}{N(\Sigma X^2) - (\Sigma X)^2}$$

Now if this formula looks familiar to you, it should. The numerator is the same numerator as the Pearson's r, and denominator is the same as the left-hand side of the Pearson's r denominator without the square root. So all we have to do in calculating the slope is use the numbers we already calculated for Pearson's r:

$$b = \frac{268}{236}$$

$$b = 1.14$$

This illustrates an important point. Where the Pearson's r will never be a number greater than 1, the slope can be greater than 1.

The next step is to calculate the y intercept as follows:

$$a = \overline{Y} - (b)\overline{X}$$

The symbol $\overline{X}$ is the mean of the X scores. We calculate this by taking the sum of the X scores (28) and dividing that by the number of pairs (10). The symbol $\overline{Y}$ is the symbol for the mean of the Y scores, which we get by taking the sum of the Y scores (34) and dividing by the number of pairs (10). We already calculated b, so now we insert the numbers:

$$a = 3.4 - (1.14)(2.8)$$
$$a = 3.4 - 3.19$$
$$a = .21$$

We now have all of the numbers we need to complete the regression equation, except one. Recall that the regression equation is

$$Y' = b(X) + a$$

where b is the slope and a is the y intercept. The last thing we need is a value for the predictor variable (X). In this case, let us predict the number of desks that someone can assemble who scored a 4 on the proficiency test. So we insert a value of 4 and calculate as follows:

$$Y' = 1.14(4) + .21$$
$$Y' = 4.56 + .21$$
$$Y' = 4.77$$

What this indicates is that we can expect someone who scores a 4 on the proficiency test to assemble roughly five desks in an hour. This was to be expected, since we had two participants in the study who scored 4 on the proficiency test and they assembled four and five desks, respectively. Since there is a very strong positive relationship between score on the aptitude test and number of desks assembled, predicting between four and five makes sense. Whether this is enough desks in an hour to get the job is a decision left to the company.

THE STANDARD ERROR OF THE ESTIMATE

Now that we have predicted how many desks someone can assemble who scores a 4 on the employment test, our next task is to determine how accurate our predictions are. We do this by computing the **standard error of the estimate**. By adding the standard error of the estimate to our predictions, we are providing a more complete description of the relationship between the predictor and criterion variables.

Unless a relationship is perfectly consistent (a Pearson's r of + or –1.00), there will be some variability, which means that we cannot make perfect predictions. However, a stronger relationship (Pearson's r closer to 1) means that there is better predictive ability than a weaker relationship (Pearson's r closer to 0). The error in a prediction is the difference between the predicted Y value (Y') and the actual Y value. Some people will score closer to Y' than others. The standard error of the estimate is the average error in our predictions. The formula for the standard error of the estimate is as follows

$$Sy' = Sy\sqrt{1 - r^2}$$

where Sy is the standard deviation of the Y scores, and r^2 is the squared value of the calculated Pearson's r. We want to include the variability in the Y scores because greater variability in the criterion variable means less predictive ability. The formula for the standard deviation of the Y scores is

$$Sy = \sqrt{\frac{\Sigma Y^2 - \frac{(\Sigma Y)^2}{N}}{N}}$$

This is the same computational formula we used in Chapter 5 to calculate the standard deviation. Here we are simply substituting the Y values for the X values. Fortunately, by calculating the Pearson's *r* we already have the values we need to solve for *Sy*:

$$Sy = \sqrt{\frac{\Sigma Y^2 - \frac{(\Sigma Y)^2}{N}}{N}}$$

Step 1: $Sy = \sqrt{\dfrac{152 - \dfrac{(34)^2}{10}}{10}}$

Step 2: $Sy = \sqrt{\dfrac{152 - \dfrac{1156}{10}}{10}}$

Step 3: $Sy = \sqrt{\dfrac{152 - 115.6}{10}}$

Step 4: $Sy = \sqrt{\dfrac{36.4}{10}}$

Step 5: $Sy = \sqrt{3.64}$

Step 6: $Sy = 1.91$

Next, we take the value of the standard deviation of the Y scores and plug it in to the formula for the standard error of the estimate, adding the Pearson's *r* value that we already calculated.

$$Sy' = Sy\sqrt{1 - r^2}$$

Step 1: $Sy' = 1.91\sqrt{1 - .91^2}$

$$\text{Step 2: } Sy' = 1.91\sqrt{1 - .83}$$

$$\text{Step 3: } Sy' = 1.91\sqrt{.17}$$

$$\text{Step 4: } Sy' = (1.91)(.41)$$

$$\text{Step 5: } Sy' = .78$$

This means that in our example we would predict that someone who scored a 4 on the proficiency test would assemble 4.77 desks in an hour, and our predictions will be off by an average of .78. We know that while there is a very strong relationship between someone's score on the proficiency test and the number of desks they can assemble, it is not a perfect relationship. So there will be some error when we predict the number of desks someone can assemble based on their score on the proficiency test. The standard error of the estimate allows us to quantify the amount of error.

THE PROPORTION OF VARIANCE ACOUNTED FOR

The **proportion of variance accounted for** is the improvement in the accuracy of our predictions when we use a relationship to predict Y scores. The differences (variance) among scores is accounted for by knowing the strength of a relationship.

Imagine that there are 10 people who are about to make desks (going back our desk-making example) and you were asked to predict how many desks each person will make in an hour. You would have nothing on which to base your answer so your predictions would be (1) totally random and (2) rife with error (most likely).

Now let us say that you were asked to make the same prediction, but this time you are given the score of each person's proficiency test, and you are also told that there is a strong positive relationship between a person's score on the proficiency test and how many desks they can assemble in an hour. You will guess a higher number of desks made for people with higher proficiency test scores, and a lower number of desks made for people with lower proficiency test scores. In short, it is better to know this relationship than to not know the relationship.

With the proportion of variance accounted for, we can statistically account for variability. The formula for the proportion of variance accounted for is very simple; we square the Pearson's r statistic:

$$\text{Proportion of Variance Accounted For} = r^2$$

For our desk-making example, Pearson's *r* is .91. Squaring this (.91²) means that we can account for roughly 83% (82.81) of the variance in desks made if we know a person's score on the proficiency test. About 83% of the variance in the number of desks made is determined by the variance in the proficiency test scores[1]. Another way to think of this is that we improve our predictive ability by roughly 83 percent when we use a person's score on the proficiency test to predict how many desks they can make in an hour compared to not using the relationship.

The **proportion of variance not accounted for** tells us how much variability in the number of desks made is not accounted for by the relationship with scores on the aptitude test. The formula for the proportion of variance not accounted for is:

$$\text{Proportion of Variance Not Accounted For} = 1 - r^2$$

For our example, roughly 17% (17.19) of the variability in the number of desks made has nothing to do with scores on the proficiency test. Why someone makes more (or fewer) desks than someone else can be due to a number of things like work ethic or prior experience assembling desks, and these factors account for about 17% of the difference in desks made.

The proportion of variance accounted for is a very important statistic in the Behavioral Sciences because in the Behavioral Sciences we want to predict behavior. The greater the proportion of variance accounted for, the more important the relationship.

RESTRICTION OF RANGE

When we make predictions using a Pearson's r correlation coefficient, we are limited by something called the restriction of range. In our example, the scores on the proficiency test (*X*) ranged from a low of 1 to a high of 5. The restriction of range means that if want to make a prediction based on this data, we have to choose a number for the given *X* that is between 1 and 5. We are restricted by this particular range because while the Pearson's r established that this is a positive relationship (+.91), it is possible that this is actually a curvilinear relationship, and  at a certain point beyond 5 as X keeps increasing, Y would decrease.

---

[1] The proportion of variance accounted for is also called the coefficient of determination.

CHAPTER SUMMARY

While a correlation coefficient describes the relationship between variables, linear regression is the statistical procedure we use to make predictions based on that relationship. We predict a criterion variable (Y) from a known predictor variable (X). This is done by calculating the slope, which describes the direction of the relationship, then the y intercept, which is where the regression line crosses the y axis. We then put those values into the regression equation. We can also determine how close our predictions are by calculating the standard error of the estimate, as well as the proportion of variance accounted for in the prediction.

FORMULAS FOR CHAPTER 8

Pearson's r:

$$r = \frac{N(\Sigma XY) - (\Sigma X)(\Sigma Y)}{\sqrt{\left[N(\Sigma X^2) - (\Sigma X)^2\right]\left[N(\Sigma Y^2) - (\Sigma Y)^2\right]}}$$

Slope: $b = \dfrac{N(\Sigma XY) - (\Sigma X)(\Sigma Y)}{N(\Sigma X^2) - (\Sigma X)^2}$

y intercept: $a = \overline{Y} - (b)\overline{X}$

Regression Equation: $Y' = b(X) + a$

Standard Deviation of Y Scores: $Sy = \sqrt{\dfrac{\Sigma Y^2 - \dfrac{(\Sigma Y)^2}{N}}{N}}$

Standard Error of the Estimate: $Sy' = Sy\sqrt{1 - r^2}$

Proportion of Variance Accounted For $= r^2$

Proportion of Variance Not Accounted For $= 1 - r^2$

1. For practice problem #1 in Chapter 7, use the linear regression equation to predict a quiz score for a student who studies for 4 hours. Include in your answer the standard error of the estimate as well as the proportion of variance accounted for. Round to two decimal places for all of your answers.

2. For practice problem #3 in Chapter 7, use the linear regression equation to predict a grade for a student who misses five classes. Include in your answer the standard error of the estimate as well as the proportion of variance accounted for. Round to two decimal places for all of your answers.

3. For practice problem #4 in Chapter 7, use the linear regression equation to predict a satisfaction score for a student who makes seven mistakes. Include in your answer the standard error of the estimate as well as the proportion of variance accounted for. Round to two decimal places for all of your answers.

4. A police department wants to know if the number of absences from duty in a year (Y) can be predicted from a recruit's score on an examination of mental burnout (X). First, compute Pearson's r to determine the direction and strength of this relationship. Then use the linear regression equation to predict the number of absences for a recruit who scores a 6 on the exam. Include in your answer the standard error of the estimate as well as the proportion of variance accounted for. Round to two decimal places for all of your answers.

Score on Exam of Burnout (X)	Absences in a Year (Y)
3	4
2	3
2	4
4	5
4	6
5	6
7	8
8	9
8	9
7	9

© ibreakstock/Shutterstock.com

INTRODUCTION

Question: If you flip a coin, what is the probability that it will land on heads? Second question: If you are taking a multiple-choice examination and there are five possible answers, what is the probability that taking a random guess will result in the correct answer? If you answered .50 for the first question and .20 for the second, then you already have an understanding of probability.

Now you might have answered 50% for the first question and 1 in 5 for the second, and while you would have been correct that is not how we answer in terms of probability. Probability is always expressed as a decimal. Chance is expressed as a percentage (there is a 40% chance of rain tomorrow) and odds are expressed as a ratio (the odds of someone winning the lottery are 1 in 175 million, or thereabouts), but the probability of an event happening is the number of times that event occurs divided by the number of times it can occur.

If this sounds familiar to you it should, since this is the same logic behind proportions, which we discussed in Chapters 1 and 6. The difference now is that we are using the same logic of proportions and applying it to probability.

Recall from Chapter 1 when I said that 9 out of 20 students in my class grew up outside of Pennsylvania. We said that the proportion of students in that class from outside of Pennsylvania is .45 (9/20). Another way to look at it is that the probability that one of my students in that class grew up outside of Pennsylvania is .45.

In this chapter, we will explore probability to determine the probability of an event occurring. We will also use probability to determine whether a sample represents a population, which means that this chapter is where we begin our journey into the world of inferential statistics.

DESCRIPTIVE AND INFERENTIAL STATISTICS

If you recall from Chapter 1, **descriptive** statistics are used to organize, summarize, and describe data in a sample. As we discussed in Chapter 3, all data come to us in raw (unorganized) form. Our job as researchers and statisticians is to take that raw data and organize it in such a way that it makes sense, so that information can be obtained from it easily. For example, if you want to know the

most frequently occurring score in a data set we use the mode, and if you want to know about the consistency of the data we use the standard deviation.

In Chapter 1 we also discussed **inferential** statistics, which are procedures used when organizing and describing data in a population. Since populations are usually too large to study in and of themselves, we take a sample from that population and infer the results of the sample back to the population. However, this only works if we have a representative sample. We use inferential statistics to determine whether the sample data accurately reflect the data we would find if we could study the entire population.

SAMPLING

There are different types of samples, but when we want to study a group of people and apply those findings back to the population from which they were drawn we want to use a **probability sample.** In a probability sample, every member of a population has an equal and known chance of being selected for the sample. Probability samples are unbiased, so one person does not have a different chance of being selected than anyone else.

Let us say there is a population of 10,000 people that we want to know more about but your budget only allows for you to study (through a survey) 2,000 of them. If we divide our sample size by the size of the population (in this case 2,000/10,000), we know the probability of any member of the population being selected for the sample (in this case it is .20). This helps ensure that our sample should be representative of the population, so we can be comfortable that when we draw conclusions about the sample we would draw the same conclusions about the population.

Probability samples ensure representativeness by employing **random samples.** A random sample means that one person being chosen for the sample over someone else is not controlled by the researcher. One way to think about this is to take all 10,000 names in our population and put them in a hat (it would have to be a really big hat, but bear with me). For our sample to be random, the researcher would mix up the names in the hat, close their eyes, and draw 2,000 names. This might not sound scientific, but it is because the researcher is not influencing the sampling process in any way. In practical terms, we do not draw names out of a hat. Instead we use websites that generate a set of random numbers, like randomizer.org.

Try it yourself and go to randomizer.org (it is free). Enter 1 for how many sets of numbers you want to generate. Then enter 2,000 (sample size) for numbers per set, and 1 to 10,000 for the range of numbers (population size). After you click on "randomize now," you will have a set of 2,000 random numbers between 1 and 10,000. This is the scientific equivalent of blindly drawing names out of a hat.

What makes this work is **probability theory.** Probability theory says that, over the long run, there is a statistical order to things. Over the long run is important because probability is not based on one-time events. This is illustrated by something called "the gambler's fallacy."

A roulette wheel has 37 spaces into which the ball can fall. Eighteen of these are red, 18 are black, and one is green. This makes the probability of the ball falling into one of the red spaces .49 (after rounding). Imagine a gambler at a casino who is watching the roulette wheel and they observe a pattern: on every third spin the ball falls into a black space after falling into red twice (red–red–black). This happens for nine spins, and so on the tenth spin they put all of their money on red because, well, it just has to happen. No it does not, and this is an example of the gambler's fallacy. Regardless of patterns the gambler thinks they see, the probability of the ball falling into one of the red (or black, for that matter) spaces is still .49. The gambler's fallacy is not realizing that an event happening (or not) in a limited instance does not alter its probability over the long run.

This is why when we employ probability sampling we rely on large sample sizes, relative to the size of the population. Large sample sizes are the statistical equivalent of "over the long run." However, a large sample does not mean that it will be representative of the population if every member of that population does not have an equal and known chance of being selected.

People who engage in political polling to predict elections use sampling techniques that help ensure representativeness. In 1936, President Franklin Delano Roosevelt (FDR) was running for reelection against Alf Landon. A magazine, *Literary Digest,* polled its subscribers in order to predict who would win the election. They sent surveys to its 2,000,000 subscribers and the results showed that Landon was going to beat FDR handily.

A man named George Gallup read these results and thought that they could not be true. FDR was a very popular president, based partly on his implementation of Social Security and unemployment benefits. So Gallup commissioned his own poll and surveyed 50,000 people. This is still a very large sample, but nowhere near 2,000,000. Gallup tabulated his results and agreed that the election would be won handily, but not by Landon. FDR went on to win one of the most lopsided presidential elections in United States history (523 electoral votes to Landon's 8). The important difference is that the *Literary Digest* poll was not random—only people who subscribed to their magazine had a chance of being selected for the sample. Gallup's poll was done randomly and, hence, more representative of the population (the American voter).

While employing the proper sampling techniques is crucial to having a sample that represents the population, nothing guarantees representativeness, which is one of the uses of inferential statistics. We can use inferential statistics to determine whether a sample accurately represents the population from which it was drawn. This is done by comparing the sample statistic to the population parameter. A **sample statistic** is a summary description of a given variable in a sample, and a **population parameter** is a summary description of a given variable in a population. A sample is representative of a population when there is statistical equivalence between the statistic and parameter. For example, if the mean age of a population of college students is 20.5 years, and the mean age for the sample is 20.9 years, we would have statistical equivalence. Note that the sample and parameter do not have to be the same number, but they have to be close enough to say that the sample represents the population.

We use inferential statistics and probability to decide whether what we find in the sample we would also find if we could study the entire population. In other words, whether we have statistical equivalence. We will get to that, but first let us explore probability.

PROBABILITY

We use probability to determine random, or uncontrolled, events. This means that something will happen, or not happen, by pure chance. What we do with probability is apply mathematical thinking to describe this chance. We do this by looking at how often something does, or does not, happen over the long run.

When earlier I asked the probability of flipping a coin and its landing on heads, we said that the answer was .50, but we know this because after

© SutidaS/Shutterstock.com

many flips (over the long run) the number of times a coin lands on heads and the number of times it lands on tails tends to even itself out to .50.

One year I had a student who disagreed with everything everyone in class said. When I asked the class what the probability of a coin flip landing on heads was, a student raised her hand and said 50%. (She should have said .50, but this was at the beginning of the probability lecture.) Well, the student who was always disagreeing took a quarter out of his pocket, flipped it 10 times, and got 7 heads. He then looked smugly at the student who said 50% and went to put away the quarter.

I asked him not to put the quarter away and to flip it 10 more times. The .70 heads after 10 flips (7 heads out of 10 flips) became .55 heads after 20 flips (the next 10 flips produced 4 heads). This student was ignoring an important aspect of probability. We base probability on paying attention to how many times things happen over the long run.

Theoretical and Empirical Probabilities

Two types of probability distributions are utilized: theoretical and empirical. **Empirical probabilities** are based on experience and observations (recall from Chapter 2 that empirical means "originating in or based on observation or experience"). For example, if we were to predict the number of traffic accidents in the country next year, we would have to base that estimate on the number of traffic accidents in the country over the past few years. We would collect that data, and from what we observe describe the probability of the number of traffic accidents next year.

Theoretical probabilities, on the other hand, are based on logic, independent of prior experience. If we were to roll a single die, what is the probability of getting a six? We know that the answer is .167 (1/6), and we can state the probability without having to collect data (in other words, toss a die repeatedly), because we know the six sides of the die each have an equal probability of landing on any of the six sides.

Independent and Dependent Outcomes

If the probability of an event occurring is the same regardless of another event occurring, we say that the events are **independent** of each other. A way to explain this is **sampling with replacement**. Let us say that we have a deck of cards and we want to draw a two. We shuffle the deck, and randomly draw ten cards. Since there are four two's in a deck of cards, and 52 cards in a deck (removing the Jokers), the probability of obtaining a two is .0769230769 (4/52).

Let us say we draw ten cards and do not get a two. We put the 10 cards drawn back into the deck, shuffle, and again draw 10 cards. The probability of getting a two is still .0769230769 because by putting the cards back in the deck, the total number of cards for each draw (52) did not change. This illustrates that when we sample with replacement (put the drawn cards back in the deck), we have independent events. The probability of drawing a two the second time is not influenced by the fact that we did not draw a two the first time.

However, sometimes events are dependent. Using the aforementioned example, we know that the probability of drawing a two from a deck of 52 cards is .0769230769. We shuffle the deck, draw 10 cards, and do not get a two. This time we do not put back the drawn cards; we **sample without replacement**. We reshuffle the cards and draw another ten, but now the probability of getting a two is .0952380952 because the deck of cards now has 42 cards instead of 52 (4/42). The probability increases because drawing a two the second time is **dependent** on the fact that we did not draw a two the first time. Sampling without replacement gives us dependent events.

Probability and Proportions

As we have discussed previously, there are proportions in the normal curve. For instance, we know that in a normal distribution .3413 of the scores will fall between the mean and one standard deviation above the mean. Another way of looking at this is that the probability of a score falling between the mean and one standard deviation above the mean is .3413. The proportion of scores in the normal curve is also the probability of those scores. In Chapter 6, we used the z scores to determine the proportion of scores that fall between two different scores in a normal distribution, and we will be applying that same logic here, but with a twist. The twist is instead of comparing raw scores and sample means, we are going to compare sample means and population means, since we are now using inferential statistics.

Sampling Distribution of Means

We use inferential statistics to infer from a sample to a population, but statistics can vary from one sample to the next. If we drew 20 different samples from the same population and measured the mean on some variable (let us say, age), it is possible that we could end up with 20 different sample means. How would we know that our statistic was representative of the parameter? We do this by using a theoretical probability distribution that represents a statistic for all possible samples given from a population. This theoretical probability (and it is strictly theoretical in nature) distribution is called the **sampling distribution of means.**

Let us say we have a population of college students and from it we draw a random sample of 100 ($n = 100$) and record the mean (statistic) for age. We then do this again and again (drawing samples of 100) until all of the samples have been drawn from that population, recording the statistic each time. We would then plot all of these means in a frequency distribution, but since we are using sample means the distribution is called the sampling distribution of means. This is a theoretical distribution because it is imagined; we do not actually draw sample after sample. From this, we can envision important characteristics about sampling distributions, thanks to something called the Central Limit Theorem.

The Central Limit Theorem

The **Central Limit Theorem** states the following:

1. As the size of a sample (n) increases, the shape of the sampling distribution of means more closely resembles the shape of the normal distribution. This means that even if the population of raw scores from which the sample is drawn is skewed, if the sample size is large enough, the sampling distribution will be normal.
2. The mean of the sampling distribution will equal the mean of the population ($\overline{X} = \mu$).
3. The standard deviation will equal the standard deviation divided by the square root of the sample size. This is called the **standard error of the mean** ($\sigma_{\overline{X}} = \dfrac{\sigma}{\sqrt{n}}$).

The word error is used because while a sample mean is expected to be close to the population mean, it is not expected to be exact. The standard error of the mean is how much a sample mean is expected to be different from the population mean. Another reason we like large sample sizes is that as the sample size increases, the standard error decreases. For example:

$$\sigma = 10 \quad n = 50 \qquad\qquad\qquad \sigma = 10 \quad n = 100$$

$$\sigma_{\overline{X}} = \frac{\sigma}{\sqrt{n}} = \frac{10}{\sqrt{50}} = \frac{10}{7.07} = 1.41 \qquad \sigma_{\overline{X}} = \frac{\sigma}{\sqrt{n}} = \frac{10}{\sqrt{100}} = \frac{10}{10.00} = 1.00$$

$$\sigma = 10 \quad n = 200$$

$$\sigma_{\overline{X}} = \frac{\sigma}{\sqrt{n}} = \frac{10}{\sqrt{200}} = \frac{10}{14.14} = .71$$

Larger samples tend to be a more accurate representation of the population because of the reduced error.

Probabilities of Sample Means

In Chapter 6, we used the following formula for z scores to determine the proportion of scores in a sample:

$$z = \frac{X - \overline{X}}{S}$$

where X is the raw score, $\overline{X}$ is the sample mean, and S is the sample standard deviation. For probabilities, the formula is similar, but there are significant differences because we are now using z scores to determine the probability of sample means in a population. The formula we use for this is as follows:

$$z = \frac{\overline{X} - \mu}{\sigma_{\overline{X}}}$$

where $\overline{X}$ is the sample mean, μ is the population mean, and $\sigma_{\overline{X}}$ is the standard error of the mean. Let us apply this formula to determine the probability of selecting a random sample of test scores.

We have a population of students at a college who take an exam in their senior year to determine their basic knowledge in their chosen major. In this example, let us say that the population mean (μ) is 75 (out of 100), with a population standard deviation (σ) of 15. What we want to determine is the probability of randomly selecting 25 students ($n = 25$) with a sample mean ($\overline{X}$) between 75 and 78.

First we have to determine the standard error of the mean $\left(\dfrac{\sigma}{\sqrt{n}} \right)$

$$\frac{15}{\sqrt{25}} = \frac{15}{5} = 3$$

Then we apply the z formula as follows:

$$\frac{78 - 75}{3} = \frac{3}{3} = 1$$

We have calculated a z score of 1, but that is not the probability of selecting a sample of 25 students with a mean test score between 75 and 78. What the z score does is allow us to determine the probability by referring to the proportions of scores under the normal distribution. Recall from Chapter 6 that in a normal curve .3413 of the scores fall between the mean (a z score of 0) and a z score of 1. In this context, we say that the probability of selecting a sample of 25 scores with a mean between 75 and 78 is .3413 (remember, probabilities are expressed as decimals).

In this chapter we will revisit the z tables, but instead of determining proportions we will use them to determine the probability of sample means. Let us say we want to determine the probability of selecting a sample of 50 scores in the population discussed above that have a sample mean below 70.

First we calculate the standard error of the mean

$$\sigma_{\overline{X}} = \frac{15}{\sqrt{50}} = \frac{15}{7.07} = 2.12$$

then we can solve for z

$$z = \frac{70 - 75}{2.12} = \frac{-5}{2.12} = -2.36$$

and then we look up 2.36 in the z tables. Since the z score is negative, we are looking at the left-hand side of the normal distribution. Look at column c of the z tables because the question asked for the probability of selecting a sample below 70 (if it said between the mean and z we would look at column b) and see that the probability is .0091.

DETERMINING WHETHER A SAMPLE REPRESENTS A POPULATION

Computing probability is the basis of inferential statistics. However, we also need probability to determine if a sample accurately represents a population. We use random sampling to obtain a sample that is representative of the population. In other words, random sampling should produce a sample statistic that is statistically equivalent to an underlying population parameter.

However, nothing guarantees that a sample will be representative of a population because of the possibility of **sampling error**, and every sample taken from a population contains a degree of error. When sampling error occurs, we have a sample statistic that is not statistically equivalent to the underlying population parameter. So we use probability to determine whether a statistic is close enough to a parameter to say that the sample is representative of the population, or whether the statistic is too far away from the parameter so that the sample is not representative of the population.

Chapter 3 stated that the average weight of men in the United States is 180 lb. If we were to take a sample of the men at a college and get a sample mean of 177 lb, no one would say that our sample is not representative of the population since 177 is close to 180. Remember, the statistic and parameter do not have to be the same, just statistically equivalent. However, if we took a sample of men from a college and got a sample mean of 230 lb, then that sample would not be representative because in this case our statistic and parameter are too far apart.

We will use probability to determine whether a statistic (for example, the weight of a sample of men) is close enough to the parameter (the weight of men in the population) that the sample represents a population, or is too far from a parameter so that we would have to conclude that the sample does not represent the population. To do this, we start by calculating the z score for a sample.

In keeping with the above example, we take a random sample of 100 men at a college and their mean weight is 177 pounds ($\overline{X} = 177$). We know that the average weight of men in the population is 180 pounds ($\mu = 180$), with a standard deviation of 20 pounds ($\sigma = 20$), so we calculate the z score for that sample:

$$\sigma_{\overline{X}} = \frac{20}{\sqrt{100}} = \frac{20}{10} = 2.00 \qquad z = \frac{177 - 180}{2.00} = \frac{-3}{2.00} = -1.5$$

A sample mean of 177 pounds transforms to a z-score of −1.5. Now we need another z score to compare to the calculated z score (called the **obtained z**), which is the **critical value**. The critical value (called the **z crit**) is a z score that marks the **region of rejection** on a normal distribution.

The region of rejection contains sample means that are so far from the population mean that we **reject** that the sample represents the population. A sample mean lies in the region of rejection if its z score is beyond the critical value. On the other hand, if the obtained z score does not reach the critical value, we would accept that our sample represents the population.

But how do we know where the region of rejection starts? To answer this question, we decide on a **criterion probability**, which is the probability that defines samples as too unlikely for us to accept that they represent the population. The symbol for the criterion probability is alpha (α). In research, the industry standard for the criterion probability is .05. This means that sample means occurring 5% of the time are too unlikely to represent the population.

Since there are two tails in a normal distribution, sample means can either be too high (to the right) above the population mean or too low (to the left) below the population mean that we reject that they represent the population. This means that there are two regions of rejection on a normal curve. So to determine the region of rejection, we take the .05 criterion probability and divide it by two (for each tail), which is .025.

To determine the region of rejection, go to the z tables and find in column c .025, and find the corresponding z score. It is 1.96, so the critical value is ±1.96. This means that any obtained z score from a sample that is above +1.96 or below −1.96 falls in the region of rejection and we would reject that the sample represents the population. However, if the obtained z score does not fall in the region of rejection, then we can assume that the sample does represent the population.

Now let us compare the obtained z value for our sample (−1.5) to the critical value (±1.96). We can say that our sample represents the population because the obtained z score does not fall in the region of rejection.

Here is another example. Let us imagine that the national average for the MAT is 500 ($\mu = 500$) with a standard deviation of 100 ($\sigma = 100$). We take a sample of 25 students ($n = 25$) who took the test and their sample mean is 540 ($\overline{X} = 540$). Let us transform the sample mean into a z score, compare it to the critical value, and determine whether the sample is representative of the population:

$$\sigma_{\overline{X}} = \frac{100}{\sqrt{25}} = \frac{100}{5} = 20.00 \qquad z = \frac{540 - 500}{20.00} = \frac{40}{20.00} = 2.00$$

The obtained z score of 2.00 is higher than 1.96, so we have to conclude that our sample does not represent the population since the obtained z score falls in the region of rejection.

CHAPTER SUMMARY

Probability is our first look at inferential statistics. We use probability sampling to draw a representative sample from a population. With a representative sample, we can infer what we find in a sample back to the population from which the sample was drawn. By converting raw score sample means into z scores, we can determine the probability of an event happening in a population, as well as determine whether a sample is representative of the population from which it was drawn.

FORMULAS FOR CHAPTER 9

The Standard Error of the Mean: $\sigma_{\overline{X}} = \dfrac{\sigma}{\sqrt{n}}$

z score: $z = \dfrac{\overline{X} - \mu}{\sigma_{\overline{X}}}$

1. In a deck of 52 cards, what is the probability of randomly selecting a queen?

2. In a deck of 52 cards, what is the probability of randomly selecting a card that is a 2, 7 or 10?

3. The mean of a population of scores is 40 ($\sigma = 8$). What is the probability of selecting a sample of 20 scores with a mean below 36?

4. The mean of a population of scores is 28 ($\sigma = 12$). What is the probability of selecting a sample of 30 scores with a mean above 31?

5. The mean of a population of scores is 80 ($\sigma = 14$). What is the probability of selecting a sample of 49 scores with a mean between the population mean ($z = 0$) and 82?

6. The mean of a population of scores is 68 ($\sigma = 8$). (a) What is the probability of selecting a sample of 40 scores with a mean lower than 65? (b) What is the probability of selecting a sample of 25 with a mean above 70? (c) What is the probability selecting a sample of 20 with a mean between the population mean ($z = 0$) and 72?

7. The national mean for the FBI entrance exam is 550 ($\sigma = 81$). What is the probability of selecting a sample of 25 and getting a mean of (a) 550 and above, (b) between 550 and 572, (c) below 510, (d) between 540 and 583, and (e) below 522 and above 565?

8. We conduct research to determine whether students at a nearby college are more or less in favor of the death penalty than college students in the nation, whose mean score on the death penalty opinion survey is 17 ($\sigma = 3$) (higher numbers indicate a more favorable attitude toward the death penalty). A random sample of local college students took the survey, and their scores are as follows:

17	20	25	12	11	22	22	17	17	18	18	20

   (a) With $\alpha = .05$, what is the critical value (z_{crit})? (b) What is the obtained z value (z_{obt})? (c) Do students at the local college represent the population of college students in regard to their attitudes toward the death penalty? How do you know?

9. In the population of college students taking statistics, the mean score on a final exam was 75 ($\mu = 75$, $\sigma = 6.4$). Twenty five students who studied statistics using a new technique earned an average of 72.1 on the final exam ($\overline{X} = 72.1$). Determine whether students using the new technique represent the population of statistics students using the current technique.

   (a) With $\alpha = .05$, what is the critical value (z_{crit})? (b) What is the obtained z value (z_{obt})? (c) Should we conclude that the sample belongs to and represents the population of students using the current technique? Why or why not?

10. We conduct research to determine whether students at a local college are more or less in favor of gun control than college students in the nation, whose mean score on the gun control opinion survey is 12 ($\sigma = 3.1$) (higher numbers indicate a more favorable attitude towards gun control). A random sample of students from the local college took the survey, and their scores are as follows:

14	12	15	12	11	12	12	17	17	18

   a. With $\alpha = .05$, what is the critical value (z_{crit})?
   b. What is the obtained z value (z_{obt})?
   c. Do the students at the local college represent the population of college students in regard to their attitudes toward gun control? How do you know?

CHAPTER 10
HYPOTHESIS TESTING

© dizain/Shutterstock.com

INTRODUCTION

Hypothesis testing is a procedure that we use when estimating parameters from statistics. In this chapter, we are going to use a sample to determine the likelihood of an outcome in a population. We will utilize the *z* **test** to accomplish this. The *z* test is a parametric inferential procedure.

Parametric statistical procedures require certain assumptions about the data they analyze. While there are a number of parametric inferential procedures, they all have three assumptions in common. One is that the population of dependent scores forms a normal distribution. The second is that the data are measured at the interval/ratio level, and the third is that we have a random (unbiased) sample. We will use the *z* test in this chapter to determine if the data we analyze accurately represent a relationship in an experiment (in other words, does the independent variable really have an effect on the dependent variable?), or if we are being misled by sampling error.

THE Z TEST

As stated, the z test is a parametric inferential procedure that is used to compare a sample statistic to a
population parameter (hence, parametric procedure).
However, since there is more than one parametric
inferential procedure, how are we to know when to
use the z test as opposed to other parametric procedures? The answer lies in knowing the assumptions
of the z test. Since we already know three assumptions of all parametric inferential procedures, we
know the first three assumptions of the z test, but
the second two assumptions are unique to the z test.
If your data do not meet all five assumptions, then
you will have to use another statistical test. The
assumptions of the z test are as follows:

1. The dependent variable is measured at the interval/ratio level.
2. The dependent variable is normally distributed in the population.
3. The sample was drawn using random sampling.
4. The mean of the population of raw sores (μ) is known.
5. The standard deviation of the population of raw scores (σ) is known.

When your data meet these assumptions, you can use the z test to test your hypothesis.

If you recall from Chapter 1, a hypothesis is a tentative statement about a relationship
between two variables (independent and dependent). The reason a hypothesis is a tentative
statement is because it has yet to be tested. Here is where we test that statement.

STEPS IN HYPOTHESIS TESTING

Experimental Hypothesis

The first step is establishing the **experimental hypothesis,** which is the predicted
relationship between the independent
and dependent variables in an experiment. For example, let us say that there
is a new technique for teaching statistics
that we want to test to see if it has an
effect on scores on a statistics exam. The
experimental hypothesis is that this new
method of teaching statistics (independent
variable) will have an effect on scores on
a statistics exam (dependent variable).

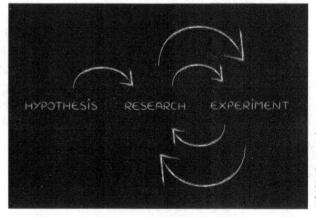

One-Tailed or Two-Tailed Test?

A relationship can be expressed in one of two ways: a **one-** or a **two-tailed** test. With a two-tailed test, we expect the independent variable to have an effect on the dependent variable, but we are unsure if scores will increase or decrease. We are predicting that scores on the dependent variable will change, but we are not predicting in which direction they will change. Scores will either increase, sending them into the right tail of the normal distribution, or they will decrease, sending them into the left tail of the normal distribution. This is why it is called a two-tailed test. We predict that scores will change, but we do not predict which way they will change.

With a one-tailed test, you do predict a direction. If we hypothesize that the new method of teaching statistics will cause scores on the statistics exam to increase, then we are predicting that scores will only go into the right tail of the normal distribution. If, for some reason, we think that the new method is ineffective and scores on the exam will decrease, then it is still a one-tailed test because we are predicting that scores will only go into the left tail of the normal distribution.

There are words that indicate a one- or two-tailed test. Words like effect, change, and alter are examples of two-tailed words. They indicate that something will happen, but not a direction. Scores will change, but we do not know into which tail of the distribution they will go. On the other hand, words like increase, improve, decrease, and decline are examples of one-tailed words because they do state a direction. Increase and improve mean that scores will go into the right tail of the distribution only, while decrease and decline mean that scores are predicted to go into the left tail of the distribution only.

Formulate the Statistical Hypotheses

The next step in hypothesis testing is to formulate our statistical hypotheses. This is done so we can apply statistical procedures to testing our hypothesis. There are two statistical hypotheses—the alternative hypothesis and the null hypothesis—and the way they are written will depend on whether you have a one-tailed or a two-tailed test.

The **alternative hypothesis** is basically a restatement of your experimental hypothesis, but this time with statistical notations. With the alternative hypothesis, we are saying that the independent variable has an effect on the dependent variable. To test this, we take a sample from a population, give the sample the independent variable, and then compare the sample and population on the dependent variable.

To test the new teaching technique for statistics, let us use a population of students who took a statistics exam after being taught using another technique. In our hypothesis, we are saying that a sample of students from this population being taught statistics using the new technique will have a different score on the exam than the population. We state this in the alternative hypothesis as follows:

$$H_A: \overline{X} \neq \mu$$

The alternative hypothesis (H, for hypothesis, with the subscript A, for alternative) is that the sample mean ($\overline{X}$) will be different (not equal to) than the population mean (μ). This is the alternative hypothesis for any two-tailed z test.

If the hypothesis is a one-tailed test, the alternative hypothesis will change slightly depending on the direction of the prediction. If we predict that the new technique improves test scores, we are saying that the sample that gets the new technique will have better scores than the population being taught with the old technique. Our alternative hypothesis will look like this

$$H_A: \overline{X} > \mu$$

This indicates that the mean scores for the sample with the new technique will be better (greater) than the mean score for the population that was taught using the old technique. If our one-tailed test predicts that the sample being taught the new technique will do worse than the population taught using the old technique, our alternative hypothesis will look like this

$$H_A: \overline{X} < \mu$$

This indicates that we are predicting that the sample will have a lower score (less than) than the population.

There is one other statistical hypothesis, and that is the null hypothesis. The **null hypothesis** states that the independent variable does not work as predicted in the experimental hypothesis. The null hypothesis is also called the hypothesis of no difference (null means nothing), and it states that the independent variable has no effect on the dependent variable. In other words, there is no relationship between the independent and dependent variable. In hypothesis testing, we always state both the alternative and null hypotheses.

For our new statistics teaching technique example, the two-tailed test states that the technique will have an effect on test scores. The null hypothesis is that the new technique has no effect on test scores; the sample that is taught with the new technique will not have different test scores than the population taught statistics using another technique. The null hypothesis for this example is

$$H_0: \overline{X} = \mu$$

The null hypothesis (H, again, for hypothesis, with a subscript zero, meaning null) is that the sample mean is not different from (is equal to) the population mean. Both of your statistical hypotheses have to account for every outcome of an experiment. In a two-tailed test, the independent variable either has an effect (scores will be different) or it does not have an effect (scores will not be different).

In a one-tailed test, predicting that the new technique improves test scores, the null hypothesis looks like this

$$H_0: \overline{X} \leq \mu$$

Here, the null hypothesis is that the students who were taught using the new technique will have lower test scores than the population, or the scores will be equal. Since you have to account for every possible outcome of an experiment, it is not enough to say that scores will only be lower for the sample. We have to include both possibilities in our null hypothesis because, again, we have to account for every possible outcome of an experiment. Stating the null hypothesis that the new technique will lower test scores is not enough, because there is the possibility that the new teaching technique will have no effect whatsoever. Conversely, if our alternative hypothesis is that the technique lowers test scores, then our null hypothesis looks like this:

$$H_0: \overline{X} \geq \mu$$

Once we know if it is a one- or a two-tailed test, and we have the correct statistical hypotheses, the next step is to conduct the z test.

CONDUCTING THE *Z* TEST

The formula for the z test is the same one that we used in Chapter 9. What we do now is set up the z test using the following steps.

Choose Alpha

Recall that our **criterion probability** sets up our region of rejection. In the last chapter, we learned that if a z score falls into the region of rejection, we reject that a sample represents a population. With hypothesis testing, if the calculated z score

(z obtained) falls in the region of rejection, we reject the null hypothesis.

The alpha (criterion) level determines the point where we conclude that there is a true difference between groups. If the probability of error is small, then we conclude that any differences we find in our experiment are real (not due to error). Most researchers set the criterion at .05, so $\alpha = .05$.

Locate the Region of Rejection

The region of rejection will be in one tail (one-tailed test) or both tails (two-tailed test) of a normal distribution.

Determine the Critical Value

The critical z value is referred to as the z_{crit}. The value for z_{crit} depends on whether you have a one- or a two-tailed test. If you have a two-tailed test, then the critical z value is ± 1.96. For a one-tailed test, the critical value is 1.645, but the direction is important. If you are predicting a decrease in scores (they will go into the left tail of the normal curve), then the critical value is −1.645. If you are predicting an increase in scores (into the right tail of a normal curve), then the z_{crit} is +1.645 (or you leave the + off, which indicates a positive number). The z_{crit} for a one-tailed test will never be ± because with a one-tailed test we are hypothesizing that scores will only go into one tail of the distribution. The critical value for a one-tailed test is different than a two tailed-test because we are taking the .05 criterion probability and putting it all in one tail or the other; we are not dividing it by two for both tails, which we do in a two tailed test.

Calculating the *Z* Score

The last step is to calculate the *z* score. Let us calculate *z* to see if using a different method of teaching statistics has an effect on exam scores. We have a population of students who took a statistics test after being taught statistics with method A. The mean population score on the test is 72 ($\mu = 72$), with a standard deviation of 10 ($\sigma = 10$). From that population we randomly sample 25 students ($n = 25$), teach them statistics using method B, and then have them take a statistics exam. Their mean score is 76. The question for us is whether 76 is different enough from 72 to say that the new method had an effect on test scores. We will determine this by calculating the *z* score for a sample mean of 76.

Since we want to know if the new method had an effect on test scores, this is a two-tailed test (we are not predicting a direction). We first state our statistical hypotheses

$$H_A: \overline{X} \neq \mu; \quad H_0: \overline{X} = \mu$$

With a criterion of .05 ($\alpha = .05$) our critical value is ±1.96 ($z_{crit} = \pm 1.96$). To first calculate the *z* score, we have to calculate the standard error of the mean

$$\sigma_{\overline{X}} = \frac{\sigma}{\sqrt{n}} = \frac{10}{\sqrt{25}} = \frac{10}{5} = 2$$

Then we can calculate the *z* score (z_{obt})

$$z = \frac{\overline{X} - \mu}{\sigma_{\overline{X}}} = \frac{76 - 72}{2} = \frac{4}{2} = 2$$

Our z_{obt} is greater than the z_{crit}, so we reject the null hypothesis since our obtained *z* falls in the region of rejection (any value equal to or greater than ±1.96). In statistics, we always test the null because it is more conservative to say that there is no difference between groups than it is to say that a difference is due to error. In inferential statistics, we assume that the null hypothesis is always true.

If we reject the null hypothesis, we then accept, or retain, the alternative hypothesis. For the current example, we would then conclude that the new method for teaching statistics has an effect on test scores, and the new technique leads to a statistically significant improvement in test scores.

Significant in statistical terms does not mean important. Rather, when we say that something is **statistically significant,** we are saying that there is a reasonable level of certainty that the results did not happen by chance. In other words, there is a high probability (in this case, .95) that this new teaching method improves test scores.

What if instead we hypothesize that a new technique for teaching statistics increases scores on a statistics test? Since this is a one-tailed test, our statistical hypotheses are as follows

$$H_A: \overline{X} > \mu; \ H_0: \overline{X} \leq \mu$$

There is also a different critical z value, which in this case is $+1.645$. It is a positive number because we are predicting an increase in scores. Had we predicted that the new technique leads to a decrease in scores, the critical z value would be -1.645.

Let us say there is another population of students taking statistics who were taught using method C. They take a statistics test and their mean score is 70, with a standard deviation of 9 ($\mu = 70, \sigma = 9$). We take a random sample of 50 ($n = 50$) students from that population, teach them statistics using method D, and have them take a statistics test. Their mean score is 72 ($\overline{X} = 72$). To find if this is a significant difference, we have to conduct the z test

$$\sigma_{\overline{X}} = \frac{\sigma}{\sqrt{n}} = \frac{9}{\sqrt{50}} = \frac{9}{7.07} = 1.27$$

$$z = \frac{\overline{X} - \mu}{\sigma_{\overline{X}}} = \frac{72 - 70}{1.27} = \frac{2}{1.27} = 1.57$$

Since our obtained z score does not fall beyond the critical z value ($+1.645$), it is not in the region of rejection, and so we retain the null hypothesis. For the current example, we would conclude that the new method for teaching statistics does not work as predicted. Yes, scores for the experimental group were higher than the population, but not by enough that we would say the difference is statistically significant. The way we communicate findings when the null hypothesis is retained is that the results of our experiment are **nonsignificant,** which simply means that no relationship was found and that any differences between the sample and population are due to chance.

ERRORS IN PREDICTION

When you are testing hypotheses by comparing samples to populations, there is always the possibility of sampling error. Due to random chance, we could have a sample that does not represent the population from which it was drawn.

Consider our two-tailed test when we sought to determine whether the new method for teaching

© Borka Kiss/Shutterstock.com

statistics has an effect on exam scores. We draw our sample from a population of students who were taught with a different method than the one we are testing. The mean score on a statistics exam for those students was 72, so our sample mean before using the new method was probably 72, or right around there, because our sample should be representative of the larger population of students from which it was randomly drawn. Teaching them statistics using a new technique resulted in a sample mean of 76 on a statistics exam. Since this was significantly higher than the population mean of 72, we concluded that it was likely that the increase was due to the new teaching technique.

However, we cannot say that the new technique caused the increased test scores because the increase could have been due to sampling error. Maybe we oversampled really intelligent students who were not representative of the population and they would have done well on the exam regardless of teaching technique. If this is the case, then the null hypothesis—that the new teaching technique does not have an effect on test scores—should have been retained, not rejected. When we reject a true null hypothesis, we have made a **Type I Error**.

We never know if we have made a Type I Error, but the probability of making one is alpha (α). By setting our criterion at .05, we are accepting a 5% chance that our rejection of the null hypothesis was incorrect. There is never 100% certainty.

One way to think about this is an analogy to the criminal justice system. In a criminal court, defendants are found guilty beyond a reasonable doubt. No one is convicted based on absolute certainty. Hypothesis testing works the same way. When we set our criterion, we are implicitly saying that there is always the possibility of sampling error. However, if there is a 5% chance that we rejected the null hypothesis when we should not have, the probability is .95 ($1 - \alpha$) that we did not make a Type I Error.

Another type of error can occur when you retain the null hypothesis when it is false. This is called a **Type II Error**. With a Type II Error, we are concluding that the independent variable does not have an effect on the dependent variable when, in fact, it does. If you look at our second example (the one-tailed test), we failed to reject the null hypothesis because the sample mean of 72 was too close to the population mean for us to say that it was significantly higher (remember we hypothesized an increase in scores).

However, this result also could have been due to sampling error. Perhaps we oversampled students who are poor at statistics who were not representative of the population and would not score higher than the average regardless of teaching technique. The probability of making a Type II Error is beta (β). The calculations for beta are beyond what we need to know at this level of statistics, but if β is the probability of making a Type II Error, then the probability of not making a Type II Error is $1 - \beta$.

To sum, in a Type I situation we either make a Type I Error, the probability of which is α, or we are correct and have avoided a Type I Error, the probability of which is $1 - \alpha$. Or we have a Type II situation where we make a Type II Error, the probability of which is β, or we are correct and have avoided a Type II Error, the probability of which is $1 - \beta$.

CHAPTER SUMMARY

Hypothesis testing is a procedure we use when estimating population parameters from sample statistics. We use parametric statistics when our data meet certain assumptions. The z test is a parametric inferential procedure we use when both the population mean and population standard deviation are known. Since random sampling does not guarantee a representative sample, we have to guard against making a Type I or a Type II error when drawing conclusions.

FORMUALS FOR CHAPTER 10

The Standard Error of the Mean: $\sigma_{\overline{X}} = \dfrac{\sigma}{\sqrt{n}}$; z-score: $z = \dfrac{\overline{X} - \mu}{\sigma_{\overline{X}}}$

1. For each of the following hypotheses, state whether it is a one- or a two-tailed test:
   a. Whether the amount of sleep during finals week for college students increases or decreases compared to the rest of the semester.
   b. Drinking alcohol leads to a decrease in reaction time while driving.
   c. Productivity of workers in a warehouse is affected by decreased lighting.
   d. Increased study time has an effect on exam grades.

2. A training is designed to increase a participant's ability concentrate. The national average for the National Concentration Exam (NCE) is 79 ($\sigma = 5$). A sample of people participate in the training and receive the following NCE scores:

90	95	82	76	80	83	84	80	88	68

   a. Is this a one-tailed or a two-tailed test?
   b. What is the alternative hypothesis?
   c. What is the null hypothesis?
   d. With $\alpha = .05$, what is the z_{crit}?
   e. What is the z value (z_{obt})?
   f. Is there a relationship between the training and scores on the NCE? If so, describe the relationship.

3. The Internal Revenue Service has claimed that the mean number of times the average U.S. citizen has cheated on their taxes in the last ten years is 5.6 ($\sigma = 1.4$). You think the number is actually higher than this, so you take a random sample of 64 citizens who pay taxes and find the reported number of times they have cheated on their taxes in the last ten years is 6.3.
   a. Is this a one-tailed or a two-tailed test?
   b. What is the alternative hypothesis?
   c. What is the null hypothesis?
   d. With $\alpha = .05$, what is the z_{crit}?
   e. What is the z value (z_{obt})?
   f. What do you conclude about the number of times people have cheated on their taxes in the last ten years compared to the national average?

4. Over a 20-year period, the average sentence given to people convicted of aggravated assault in the United States was 25.9 months ($\sigma = 6.5$). You think this might be different in

your home state so you take a random sample of 75 jurisdictions and find that the mean sentence for aggravated assault is 27.3 months.

a. Is this a one-tailed or a two-tailed test?
b. What is the alternative hypothesis?
c. What is the null hypothesis?
d. With $\alpha = .05$, what is the z_{crit}?
e. What is the z value (z_{obt})?
f. What do you conclude about the mean sentence for aggravated assault in your state compared to the national average?

5. The American Bar Association reports that the mean length of time for a juvenile court hearing in the population is 25 minutes ($\sigma = 6$). As a lawyer who practices in juvenile court, you think the average is shorter than this. You take a sample of 20 other lawyers who do juvenile court work and ask them the length of their last juvenile court hearing and get a mean of 23 minutes.

a. Is this a one-tailed or a two-tailed test?
b. What is the alternative hypothesis?
c. What is the null hypothesis?
d. With $\alpha = .05$, what is the z_{crit}?
e. What is the z value (z_{obt})?
f. What do you conclude about your assumption regarding the length of juvenile court hearings?

6. The national average for the number of patients seen in emergency rooms in one shift is 50 ($\sigma = 3$). You think it is different in the hospital where you work so you take a random sample of 16 shifts and get an average of 52 patients seen in one shift.

a. Is this a one-tailed or a two-tailed test?
b. What is the alternative hypothesis?
c. What is the null hypothesis?
d. With $\alpha = .05$, what is the z_{crit}?
e. What is the z value (z_{obt})?
f. What do you conclude about your assumption regarding the number of emergency room patients seen in one shift in your hospital compared to the national average?

CHAPTER 11

THE ONE-SAMPLE *T*-TEST

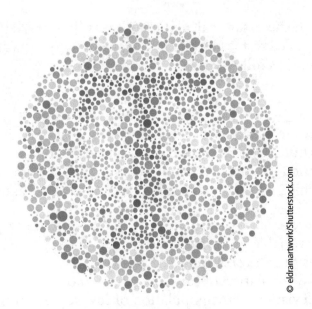

© eldramartwork/Shutterstock.com

INTRODUCTION

The *t* test is like the *z* test in that they are both parametric inferential procedures. The key difference is that, unlike the *z* test, with the *t* test we do not know the population standard deviation (σ). Because of this, the *t* test is used more often than the *z* test in Behavioral Sciences research. National averages (μ) can be found, or calculated, from any number of sources that publish national statistics (for example, The Census Bureau or the Uniform Crime Report). However, what is never (or rarely) included with the national averages is the amount of variability. With the *t* test, we will estimate the variability by calculating the estimated population variance.

THE *t* TEST

In the last chapter, we used the *z* test to determine whether a new method of teaching statistics had an effect on test scores. Let us revisit that example, but this time we are going to determine whether this new teaching technique has an effect on test scores using the *t* test.

The national average on a statistics examination is 70 ($\mu = 70$). To test whether this new teaching technique has an effect on test scores, we randomly sample ten students, teach them statistics using the new technique, and then give them the statistics exam. The ten students get the following scores:

65   70   93   78   72   99   84   75   70   89

We will engage in the necessary calculations shortly, but first let us set up our testing. Since we have a two-tailed test ("has an effect"), we state our statistical hypotheses (the alternative and the null) as follows:

$$H_A: \overline{X} \neq \mu; \ H_0: \overline{X} = \mu$$

As you can see, they are the same as the *z* test. Next we set our criterion at .05 ($\alpha = .05$). The criterion decides our region of rejection. This means that if our calculated *t* value falls in the region of rejection, we reject the null hypothesis. If it does not, we retain the null hypothesis. The last thing we do is make sure our data meet the assumptions of the *t* test:

1. The dependent variable is measured at the interval/ratio level.
2. The dependent variable is normally distributed in the population.
3. The sample was drawn using random sampling.
4. The mean of the population of raw sores (μ) is known.
5. The standard deviation of the population of raw scores (σ) is not known.

Now we can conduct the one sample *t* test.

CALCULATING THE TWO-TAILED *t* TEST

Since the population variability is unknown, we have to estimate it using the following formula:

Estimated Population Variance: $v = \dfrac{\sum X^2 - \dfrac{(\sum X)^2}{N}}{N - 1}$

It is very similar to the formula for the sample variance we saw in Chapter 5, but since we are estimating the population variability from a sample, the denominator is the **degrees of freedom**, $N - 1$ (which was discussed

in Chapter 5). To calculate the estimated population variance, we need to add up all of the test scores (X) for the students learning statistics with the new technique, and we will also have to square the value of each test score (X^2) and add them all together.

X	X^2
65	4,225
70	4,900
93	8,649
78	6,084
72	5,184
99	9,801
84	7,056
75	5,625
70	4,900
89	7,921
Sum = 795	64,345

We can now calculate the sample mean by dividing the sum of X (795) by the sample size (10), and it is 79.5. So now we have to decide whether 79.5 is significantly different enough from 70 for us to say that the new teaching technique has an effect on test scores, or whether it is close enough to 70 that it does not.

Our next step is to calculate the estimated population variance:

$$\text{Step 1: } v = \frac{64,345 - \dfrac{(795)^2}{10}}{9}$$

$$\text{Step 2: } v = \frac{64,345 - \dfrac{632,025}{10}}{9}$$

$$\text{Step 3: } v = \frac{64,345 - 63,202.5}{9}$$

$$\text{Step 4: } v = \frac{1,142.5}{9}$$

$$\text{Answer: } v = 126.94$$

The formula for the t test is:

$$t = \frac{\overline{X} - \mu}{S_{\overline{X}}}$$

This is very similar to the formula for the z test, but there is an important difference. The denominator for the z test is the standard error of the mean, but since we are estimating the variability in the population, in the t test we have to divide the numerator by the estimated standard error of the mean. The formula for this is

$$S_{\overline{X}} = \sqrt{\frac{v}{N}}$$

To calculate the estimated standard error of the mean, we simply divide the estimated population variance by the sample size, and then take the square root:

$$\text{Step 1: } S_{\overline{X}} = \sqrt{\frac{126.94}{10}}$$

$$\text{Step 2: } S_{\overline{X}} = \sqrt{12.69}$$

$$\text{Answer: } S_{\overline{X}} = 3.56$$

Now, we are ready to solve for t:

$$t = \frac{\overline{X} - \mu}{S_{\overline{X}}}$$

$$\text{Step 1: } t = \frac{79.5 - 70}{3.56}$$

$$\text{Step 2: } t = \frac{9.5}{3.56}$$

$$\text{Answer: } t = 2.67$$

Using the t Tables

The next step in determining if the new teaching technique has an effect on exam scores is to decide whether the obtained t value (t_{obt}) is significant. We do this by comparing t_{obt} to a critical value of t (t_{crit}), which can be found in the t tables.

With the z test, our critical values were determined solely by whether it is a one- or two-tailed test. However, we cannot do this with the t test because we do not know the variability of the t distribution. The **t distribution** is the distribution of all possible values of t. Since we do not know the variability of the t distribution, the critical value will change with the sample size. Since this is the case, we have to look up the critical t value using the t tables, which can be found in Appendix A.

There are two different t tables, depending on whether we have a one- or a two-tailed test. A portion of the t tables for a two-tailed test is reproduced in Table 11.1. We look up the critical t value by first finding the correct degrees of freedom in the column labeled *df*. There are two more columns for two different alpha levels (.05 and .01). Let us set our criterion at .05, and we will use that column to look up our critical t value. In a two-tailed test with 9 degrees of freedom and $\alpha = .05$, the critical t-value is ± 2.262. This means that any t_{obt} that is greater than 2.262 or less than -2.262 is statistically significant.

TABLE 11.1 Two-Tailed t Table

df	$\alpha = .05$	$\alpha = .01$
1	12.706	63.657
2	4.303	9.925
3	3.182	5.841
4	2.776	4.604
5	2.571	4.302
6	2.447	3.707
7	2.365	3.499
8	2.306	3.355
9	2.262	3.250

Our t_{obt} of 2.67 is greater than the t_{crit} of $+2.262$, so we reject the null hypothesis (the new technique has no effect on test scores) and conclude that the new technique for teaching statistics has a statistically significant effect on exam scores. In addition, since our t_{obt} falls in the right-hand side of the t distribution, we can also conclude that the new teaching technique is likely to increase test scores.

CALCULATING THE ONE-TAILED *t* TEST

The national average on a statistics examination is 68 ($\mu = 68$). We hypothesize that a new technique for teaching statistics will improve test scores, compared with the current method of teaching statistics. To test this hypothesis, we randomly sample ten students who were taught statistics using the current method, teach them statistics using the new technique, and then give them the statistics exam. The ten students get the following scores:

© Seriv/Shutterstock.com

58   78   80   66   67   97   73   87   58   69

Since we have a one-tailed test predicting an increase in scores (improve), we state our statistical hypotheses (the alternative and the null) as follows:

$$H_A: \overline{X} > \mu; \ H_0: \overline{X} \le \mu$$

With a criterion of .05 ($\alpha = .05$), we are ready to test our hypothesis. First, we calculate the estimated population variance:

X	X²
58	3,364
78	6,084
80	6,400
66	4,356
67	4,489
97	9,409
73	5,329
87	7,569
58	3,364
69	4,761
Sum = 733	55,125

$$v = \frac{\sum X^2 - \frac{(\sum X)^2}{N}}{N - 1}$$

$$\text{Step 1: } v = \frac{55,125 - \frac{(733)^2}{10}}{9}$$

$$\text{Step 2: } v = \frac{55,125 - \frac{537289}{10}}{9}$$

$$\text{Step 3: } v = \frac{55,125 - 53728.9}{9}$$

$$\text{Step 4: } v = \frac{1,396.1}{9}$$

Answer: $v = 155.12$

Next we calculate the estimated standard error of the mean:

$$S_{\overline{X}} = \sqrt{\frac{v}{N}}$$

Step 1: $S_{\overline{X}} = \sqrt{\dfrac{155.12}{10}}$

Step 2: $S_{\overline{X}} = \sqrt{15.51}$

Answer: $S_{\overline{X}} = 3.94$

Finally, we can calculate the *t* test:

$$t = \frac{\overline{X} - \mu}{S_{\overline{X}}}$$

Step 1: $t = \dfrac{73.3 - 68}{3.94}$

Step 2: $t = \dfrac{5.3}{3.94}$

Answer: $t = 1.35$

We then have to compare our t_{obt} to the t_{crit}, which is found in the *t* tables. A portion of the *t* tables for a one-tailed test is reproduced in Table 11.2. Our t_{obt} of 1.35 is less than the t_{crit} of +1.833, so we retain the null hypothesis and conclude that the new technique has no effect on test scores. Even though the sample mean of 73.3 is higher than the population mean of 68, it is not significantly higher, and the higher test average for the sample cannot be attributed to the new teaching technique.

TABLE 11.2 One-Tailed t Table

df	$\alpha = .05$	$\alpha = .01$
1	6.314	31.821
2	2.920	6.965
3	2.353	4.541
4	2.132	3.747
5	2.015	3.365
6	1.943	3.143
7	1.895	2.998
8	1.860	2.896
9	1.833	2.821

THE CONFIDENCE INTERVAL

If we reject the null hypothesis, the next (and final) step in the one sample *t* test is to compute a confidence interval. Remember that the *t* test is an inferential statistic, so we are inferring from our sample to the population. If we have a representative sample, we would expect to find the same results if we could test the entire population. There are two ways to do this.

One way is to use a **point estimation**, which is when we use the sample mean in a significant *t* test to infer to the entire population. In our first example, we calculated an examination sample mean of 79.5 for students who were taught statistics using the new method. In a point estimation, we would then estimate that everyone in the population taught statistics using the new method would also earn 79.5 on the exam.

However there are two problems with this. One is that we would not expect everyone in the population to score a 79.5. As a matter of fact, no one in the sample scored 79.5 on the statistics exam. The second problem is that a point estimation ignores the variability in the scores, which in our sample ranged from 99 to 65. To avoid these problems, we use something called an interval estimation.

An **interval estimation** is a range of scores that we would expect the population mean to fall into. You might be familiar with this as it is sometimes referred to as the margin of error. The news might report that a candidate running for office is polling at 70% of the vote, with a margin of error of ±3%. This means that 70% of the voters sampled indicated that they are going to vote for that candidate, but reporters draw conclusions about the entire population of voters (which you can do with a representative sample). By saying that the margin of error is ±3%, the reporters are indicating that if the entire population of voters were sampled, they would find that between 67% and 73% of the voters would vote for that candidate. With an interval estimation, we state with probability that our statistic is within a certain distance of the parameter. We state with a degree of confidence where the parameter would fall if we could test the entire population.

We are going to compute the 95% confidence interval, so called because we set our criterion at .05. The formula for the 95% confidence interval is

$$(S_{\overline{X}})(-t_{\text{crit}}) + \overline{X} < \mu < (S_{\overline{X}})(+t_{\text{crit}}) + \overline{X}$$

For the lower end of the range, we take the calculated estimated standard error of the mean and multiply it by the negative critical *t* value (which we found in the *t* tables), then add the number to the sample mean. For the higher end of the range, we take the estimated standard error of the mean, multiply it by the positive critical *t* value, and add that to the sample mean. When calculating the 95% confidence interval, we always use the critical value for a two-tailed test, even if we calculated the one-tailed *t*-test.

For our first problem in this chapter, we calculated a sample mean of 79.5, which we found was significantly different from the population mean of 70 so the 95% confidence interval is

$$(S_{\overline{X}})(-t_{\text{crit}}) + \overline{X} < \mu < (S_{\overline{X}})(+t_{\text{crit}}) + \overline{X}$$

Step 1: $(3.56)(-2.262) + 79.5 < \mu < (3.56)(2.262) + 79.5$

Step 2: $-8.05 + 79.5 < \mu < 8.05 + 79.5$

Step 3: $71.45 < \mu < 87.55$

For this example, we are 95% confident that if the entire population of students taking statistics were taught using the new technique, they would score between 71.45 and 87.55 on the statistics exam.

We only do this if we have a significant one sample t test, meaning we rejected the null hypothesis. In our one-tailed t test example, we did not reject the null hypothesis, which indicates that the sample mean for students being taught statistics using the new method (73.3) was not significantly different from the population mean of students who were taught statistics using the current method (68). This is because there is not a statistically significant difference between the two methods, even though they are different numbers. The difference we see here is due to chance, and not the new teaching technique.

CHAPTER SUMMARY

The t test is a parametric inferential procedure like the z test where we compare samples to populations. However, unlike the z test, with the t test we do not know the population standard deviation. This means that the variability in the population has to be estimated by calculating the estimated standard error of the mean. With significant results, we calculate a confidence interval to estimate from our sample to a population.

FORMULAS FOR CHAPTER 11

The Estimated Population Variance $= v = \dfrac{\sum X^2 - \dfrac{(\sum X)^2}{N}}{N-1}$

The Estimated Standard Error of the Mean $= S_{\overline{X}} = \sqrt{\dfrac{v}{N}}$

The One Sample t Test $= t = \dfrac{\overline{X} - \mu}{S_{\overline{X}}}$

95% Confidence Interval $= (S_{\overline{X}})(-t_{\text{crit}}) + \overline{X} < \mu < (S_{\overline{X}})(+t_{\text{crit}}) + \overline{X}$

1. For each of the obtained t values given below, determine the critical t value and if the null hypothesis (H_0) should be rejected.

df	α	One or Two Tailed Test	t_{obt}	t_{crit}	Reject H_0 (Yes/No)
11	.05	One-tailed (increase)	2.37	_____	_____
15	.05	One-tailed (decrease)	−2.98	_____	_____
10	.01	Two-tailed	3.17	_____	_____
21	.05	Two-tailed	2.55	_____	_____
9	.01	One-tailed (increase)	2.22	_____	_____
22	.05	Two-tailed	2.11	_____	_____
7	.05	Two-tailed	2.04	_____	_____
18	.05	One-tailed (decrease)	−1.12	_____	_____

2. Researchers want to test if a new Anatomy and Physiology textbook has an effect on grades on a 10-point quiz. The population of students using the current textbook averaged 6.9 on the quiz ($\mu = 6.9$). A sample of students using the new textbook are given the quiz and get the following scores:

7	7	9	8	7	10	8	7	7	9

a. Is this a one- or a two-tailed test?
b. What are H_A and H_0?
c. With $\alpha = .05$, what is the critical t value (t_{crit})?
d. What is the obtained t value (t_{obt})?
e. Does the new textbook have an effect on quiz scores? If so, what is it?
f. If you rejected the null hypothesis, what is the 95% confidence interval?

3. Listening to music with earbuds lowers your ability to hear. The national mean in a hearing exam of people who listen to music without earbuds is 16.4 (μ = 16.4) (higher scores indicate better hearing). A sample of people who listen to music with earbuds takes the hearing exam and gets the following scores:

15	13	18	16	15	17	16	14	17	18	11	13

   a. Is this a one- or a two-tailed test?
   b. What are H_A and H_0?
   c. With α = .05, what is the critical t value (t_{crit})?
   d. What is the obtained t value (t_{obt})?
   e. Does listening to music with earbuds lower your ability to hear? How do you know?
   f. If you rejected the null hypothesis, what is the 95% confidence interval?

4. A study conducted by the Research Institute of America has concluded that the average number of hours inmates at state correctional facilities spend in their cells during a day is 14 (μ = 14). You obtain a sample of inmates in your state to see if the average time spent in a cell differs from the national average:

16	21	14	13	22	15	18	19
14	9	10	21	22	15	13	

   a. Is this a one- or a two-tailed test?
   b. What are H_A and H_0?
   c. With α = .05, what is the critical t value (t_{crit})?
   d. What is the obtained t value (t_{obt})?
   e. Does the amount of time prisoners in your state spend in their cell differ from the national average? If so, how do you know?
   f. If you rejected the null hypothesis, what is the 95% confidence interval?

5. The mean score for golfers at the local country club is 87 (μ = 87). The club's owner instructs the manager to post warning signs at the sand traps and ponds to see if it will improve scores. After posting the signs, the manager randomly samples 15 golfers who have the following scores:

86	86	85	90	86	77	78	83
83	90	90	83	83	86	80	

   a. Is this a one- or a two-tailed test?
   b. What are H_A and H_0? (Note: golf scores improve if they decrease.)
   c. With α = .05, what is the critical t value (t_{crit})?
   d. What is the obtained t value (t_{obt})?
   e. Did posting warning signs improve golf scores? If so, how do you know?
   f. If you rejected the null hypothesis, what is the 95% confidence interval?

The Independent Samples *t*-Test

Testing

© Kachka/Shutterstock.com

INTRODUCTION

In this chapter we are going to move from comparing a sample to a population to comparing two samples to each other. The independent samples *t* test is used when we compare two different groups to each other.

THE INDEPEDENT SAMPLES *t* TEST

In Chapter 2 we discussed the parts of an experiment. If you recall, they are experimental and control groups, pre- and posttesting, and independent and dependent variables. The experimental group is given the independent variable, the control group is not, and then both groups are tested for any differences between them. If there are any differences, they can be attributed to the effect of the independent variable. The way we measure to see if

© Magi Bagi/Shutterstock.com

there are differences between the two groups is by conducting an independent samples *t* test, given that your data meet the assumptions of running such a test, which are as follows:

1. Random sampling.
2. The dependent variable is measured at the interval/ratio level.
3. The dependent variable is normally distributed in the population.
4. The populations from which the samples are taken have homogeneous variances (the variances of the populations being represented are equal).
5. Members were assigned to groups randomly (without any bias).
6. The size of each group should be roughly equal.

Conditions five and six require some elaboration. The reason this test is called an independent samples *t*-test is that the groups are independent of each other. This means that membership in one group is not dependent on membership in the other. In other words, being in one group means that you cannot be in the other. When we assign individuals to groups, we do so randomly and without bias, so whichever group one ends up in is simply a matter of chance.

In Chapter 9 we discussed random sampling, and we likened random sampling to blindly drawing names out of hat. Not very scientific sounding, but there is no bias involved. You can liken assigning individuals to independent groups to blindly drawing names out a hat and then blindly assigning them to one of two groups, experimental and control. Therefore we guarantee that if you are in one group there is no way you can be in the other, and that there is no bias in group assignment.

Condition six, roughly equal group sizes, is trickier. There is no hard and fast rule for this, and it requires judgment on the part of the researcher. While some will say that the group sizes should not be off by more than 20, this depends on the size of the groups. If you have a sample size of 40 and one group has 28 and the other has 12, the independent samples *t* test is not appropriate, even though the difference is less than 20. However, if you have a sample of 1,040 and one group has 500 and the other has 540, then you can conduct an independent samples *t* test (which I have). The *t* test is a robust test (which we discussed in Chapter 1), which means that it can tolerate some violations of its assumptions.

CALCULATING THE TWO-TAILED INDEPENDENT SAMPLES *t* TEST

In Chapter 5 we examined the relationship between taking a pre-employment test and the number of desks a worker can assemble. In this chapter, we will put a different spin on that example.

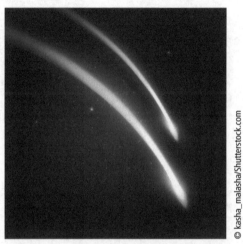

Let us say that a furniture manufacturer develops a training designed to teach people how to quickly assemble desks. The company would like to see if the training has an effect on the number of desks people can assemble, so they ask for volunteers to be part of an experiment. Twenty-three people apply to be part of the experiment, and they are randomly assigned to two groups: 11 are part of the experimental group and 12 are assigned to the control group. Since the experimental group receives the independent variable, they attend the training on how to assemble desks, and the control group does not. Then each group is asked to assemble as many desks as they can in 1 hour. To judge the effectiveness of the training, the manufacturer compares the mean number of desks made in each group to see if there is a statistically significant difference between them.

Since we are testing whether there is a difference between the groups, and not specifying a direction of that difference (has an effect), this is a two-tailed test. Our alternative hypothesis will look slightly different from the alternative hypothesis we used with the *z* test and the one-sample *t* test, because we are now comparing two sample means and not a sample mean and a population mean. Therefore, the alternative hypothesis for a two-tailed independent samples *t* test is

$$H_A: \overline{X}_1 \neq \overline{X}_2$$

which states that the mean score of group 1 ($\overline{X}_1$) will be different than the mean score of group 2 ($\overline{X}_2$). The null hypothesis for a two-tailed independent samples *t* test is

$$H_0: \overline{X}_1 = \overline{X}_2$$

which states that there will be no difference between the two groups, and hence, the training has no effect on the number of desks assembled.

The experimental group (group 1) goes through the training, while the control group (group 2) does not. Then both groups assemble desks for 1 hour, and these are the results:

Group 1						Group 2					
3	4	2	3	2		3	2	1	2	1	2
2	4	3	2	3	3	0	2	2	1	4	1

The mean number of desks assembled in an hour for the experimental group is 2.82, and the mean number of desks for the control group is 1.75. Obviously there is a difference, but we need to conduct an independent samples *t*-test to see if this difference is due to the effect of the training, or if it is due to random chance. We will set our criterion (α) at .05.

First, we have to determine the variability in each group's scores, so we will compute the estimated population variance for each group:

$$v = \frac{\sum X^2 - \frac{(\sum X)^2}{N}}{N - 1}$$

In order to do this, we will have to square each value in both of the groups, and then add them together:

Group 1		Group 2	
X	**X²**	**X**	**X²**
3	9	3	9
4	16	2	4
2	4	1	1
3	9	2	4
2	4	1	1
2	4	2	4
4	16	0	0
3	9	2	4
2	4	2	4
3	9	1	1
3	9	4	16
31	93	1	1
		21	49

$$\text{Step 1: } v_1 = \frac{93 - \frac{(31)^2}{11}}{10} \qquad v_2 = \frac{49 - \frac{(21)^2}{12}}{11}$$

$$\text{Step 2: } v_1 = \frac{93 - \frac{961}{11}}{10} \qquad v_2 = \frac{49 - \frac{441}{12}}{11}$$

$$\text{Step 3: } v_1 = \frac{93 - 87.36}{10} \qquad v_2 = \frac{49 - 36.75}{11}$$

$$\text{Step 4: } v_1 = \frac{5.64}{10} \qquad v_2 = \frac{12.25}{11}$$

$$\text{Step 5: } v_1 = .56 \qquad v_2 = 1.11$$

We need the estimated population variance for each group because the next statistic that we compute is the **pooled variance** (V_{pool}). The pooled variance is the weighted average of the sample variances, and the formula looks like this:

$$V_{pool} = \frac{(n_1 - 1)(v_1) + (n_2 - 1)(v_2)}{(n_1 - 1) + (n_2 - 1)}$$

In the numerator, we multiply the degrees of freedom for each group ($n - 1$) by the group's estimated population variance, and then add those numbers together. The denominator is the degrees of freedom for both groups added together:

$$\text{Step 1: } V_{pool} = \frac{(11 - 1)(.56) + (12 - 1)(1.11)}{(11 - 1) + (12 - 1)}$$

$$\text{Step 2: } V_{pool} = \frac{(10)(.56) + (11)(1.11)}{(10) + (11)}$$

$$\text{Step 3: } V_{pool} = \frac{5.6 + 12.21}{21}$$

$$\text{Step 4: } V_{pool} = \frac{17.81}{21}$$

$$\text{Step 5: } V_{pool} = .85$$

The weighted average of the sample variances is .85, and we need that number to enter into our next formula, which is the standard error of the mean difference $(V_{\overline{X}_1-\overline{X}_2})$. The **standard error of the mean difference** is the standard deviation of the sampling differences between the sample means. The formula for the standard error of the mean difference is

$$V_{\overline{X}_1-\overline{X}_2} = \sqrt{(V_{pool})\left(\frac{1}{n_1} + \frac{1}{n_2}\right)}$$

in which we multiply the pooled variance by 1 divided by the sample size of the first group, added to 1 divided by the sample size of the second group, and then we take the square root.

Step 1: $V_{\overline{X}_1-\overline{X}_2} = \sqrt{(.85)\left(\frac{1}{11} + \frac{1}{12}\right)}$

Step 2: $V_{\overline{X}_1-\overline{X}_2} = \sqrt{(.85)(.09 + .08)}$

Step 3: $V_{\overline{X}_1-\overline{X}_2} = \sqrt{(.85)(.17)}$

Step 4: $V_{\overline{X}_1-\overline{X}_2} = \sqrt{.14}$

Step 5: $V_{\overline{X}_1-\overline{X}_2} = .37$

The standard error of the mean difference is the denominator for the independent samples t-test, which means that we can now determine if the difference between the sample means is statistically significant. The formula for the independent samples t-test is:

$$t = \frac{\overline{X}_1 - \overline{X}_2}{V_{\overline{X}_1-\overline{X}_2}}$$

The mean for the experimental group $(\overline{X}_1)$ is 2.82 and the mean for the control group $(\overline{X}_2)$ is 1.75; so to calculate the independent samples t test, we plug the numbers into the formula

Step 1: $t = \dfrac{2.82 - 1.75}{.37}$

Step 2: $t = \dfrac{1.07}{.37}$

Step 3: $t = 2.89$

Using the *t* Tables

Recall from Chapter 11 that there are two different *t* tables, depending on whether we have a one- or a two-tailed test. A portion of the *t* tables for a two-tailed test is reproduced in Table 12.1. We look up the critical *t* value by first finding the correct degrees of freedom in the column labeled *df*. There are two more columns for two different alpha levels (.05 and .01). We set our criterion at .05 and so we will use that column to look up our critical *t* value.

TABLE 12.1 Two Tailed *t* Table

df	$\alpha = .05$	$\alpha = .01$
17	2.110	2.898
18	2.101	2.878
19	2.093	2.861
20	2.086	2.845
21	2.080	2.831
22	2.074	2.819
23	2.069	2.087
24	2.064	2.797
25	2.060	2.787
26	2.056	2.779

Since this is an independent samples *t* test comparing two groups, we use a slightly different formula for the degrees of freedom:

$$df = (n_1 - 1) + (n_2 - 1)$$

We already calculated this when we were solving for the pooled variance, so for this problem the degrees of freedom are 21. In a two-tailed test with 21 degrees of freedom and $\alpha = .05$, the critical *t* value is ±2.080. This means that any t_{obt} that is greater than 2.080 or less than -2.080 is statistically significant.

Our t_{obt} of 2.89 is greater than the t_{crit} of +2.080, so we reject the null hypothesis (the training has no effect on the time it takes to assemble desks) and conclude that the training has a statistically significant effect on desk assembly. In addition, since our t_{obt} falls in the right-hand side of the *t* distribution, we can also conclude that the new teaching technique is likely to increase the number of desks assembled.

EFFECT SIZE

When a significant relationship is found in an independent samples *t* test, the last step is calculating the **effect size** of the independent variable. A significant result shows that

© Dusit/Shutterstock.com

the independent variable (in this case, the desk assembly training) had a significant effect on the dependent variable (the number of desks assembled). The question, then, is how much of an effect?

The larger the effect size, the greater the impact of the independent variable. Not all independent variables are created equal; so even though there is an effect it does not mean that the effect will be the same in all experiments. It is worth emphasizing that effect size is only calculated when there are significant results (in other words, when the null hypothesis is rejected). If the results are nonsignificant, then the independent variable did not have an effect on the dependent variable, and so looking for an effect size makes no sense.

The Point Biserial Correlation Coefficient

One way to measure effect size is how consistently changing conditions of the independent variable changes the conditions of the dependent variable. We measure this through the proportion of variance accounted for (as discussed in Chapter 8), which we can find in a two-sample experiment using the **Point Biserial Correlation Coefficient** (r^2_{pb}). Just like the Pearson's r, this will have a range from 0 (no effect) to 1 (total effect which means 100% accuracy in predictions), but unlike the Pearson's r the Point Biserial will always be a positive number. The formula for the Point Biserial Correlation Coefficient is

$$r^2_{pb} = \frac{(t_{obt})^2}{(t_{obt})^2 + df}$$

where we square the obtained t score, and divide it by the squared obtained t score plus the degrees of freedom. For our example, the effect size of the training between the experimental and control groups is

$$\text{Step 1: } r^2_{\text{pb}} = \frac{(2.89)^2}{(2.89)^2 + 21}$$

$$\text{Step 2: } r^2_{\text{pb}} = \frac{8.35}{8.35 + 21}$$

$$\text{Step 3: } r^2_{\text{pb}} = \frac{8.35}{29.35}$$

$$\text{Step 4: } r^2_{\text{pb}} = .28$$

This means that 28% of the variability in the number of desks assembled between the experimental and control groups is due to the training. Another way to interpret this is that people who have had the training will assemble 28% more desks than people who have not have the training.

CALCULATING THE ONE-TAILED t TEST

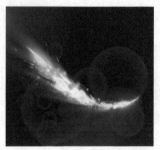

Another furniture manufacturer hears about the experiment conducted by a rival company to test their desk assembly method and decides it wants to do the same. They believe that people who take their particular training will assemble more desks than people who assemble desks without taking the training. The alternative hypothesis for a one-tailed independent samples t test predicting an increase (more desks) is

$$H_A: \overline{X}_1 > \overline{X}_2$$

which states that the mean score of group 1 ($\overline{X}_1$), the experimental group, will be higher than the mean score of group 2 ($\overline{X}_2$), the control group. The null hypothesis for a one-tailed independent samples t-test predicting an increase is

$$H_0: \overline{X}_1 \leq \overline{X}_2$$

which states that the training has no effect on desk assembly, so the group that gets the training will assemble the same number of desks, or fewer desks, than the group that does not get the training.

The experimental group (group 1) goes through the training, while the control group (group 2) does not. Then both groups assemble desks for one hour, and these are the results:

Group 1					Group 2				
3	2	2	2	3	3	2	2	2	3
4	4	3	2	3	3	3	3	2	2

The mean number of desks assembled in an hour for the experimental group is 2.8, and the mean number of desks for the control group is 2.5. Group 1 did make, on average, more desks than group 2, but we need to conduct an independent samples t test to see if this is due to the effect of the training, or if it is due to random chance. We will set our criterion (α) at .05.

First, we will compute the estimated population variance for each group:

$$v = \frac{\sum X^2 - \dfrac{(\sum X)^2}{N}}{N - 1}$$

Group 1		Group 2	
X	X^2	X	X^2
3	9	3	9
2	4	2	4
2	4	2	4
2	4	2	4
3	9	3	9
4	16	3	9
4	16	3	9
3	9	3	9
2	4	2	4
3	9	2	4
28	84	25	65

$$\text{Step 1: } v_1 = \frac{84 - \dfrac{(28)^2}{10}}{9} \qquad v_2 = \frac{65 - \dfrac{(25)^2}{10}}{9}$$

$$\text{Step 2: } v_1 = \frac{84 - \dfrac{784}{10}}{9} \qquad v_2 = \frac{65 - \dfrac{625}{10}}{9}$$

$$\text{Step 3: } v_1 = \frac{84 - 78.4}{9} \qquad v_2 = \frac{65 - 62.5}{9}$$

$$\text{Step 4: } v_1 = \frac{5.6}{9} \qquad v_2 = \frac{2.5}{9}$$

$$\text{Step 5: } v_1 = .62 \qquad v_2 = .28$$

Next we calculate the pooled variance (V_{pool}):

$$V_{\text{pool}} = \frac{(n_1 - 1)(v_1) + (n_2 - 1)(v_2)}{(n_1 - 1) + (n_2 - 1)}$$

$$\text{Step 1: } V_{\text{pool}} = \frac{(10 - 1)(.62) + (10 - 1)(.28)}{(10 - 1) + (10 - 1)}$$

$$\text{Step 2: } V_{\text{pool}} = \frac{(9)(.62) + (9)(.28)}{(9) + (9)}$$

$$\text{Step 3: } V_{\text{pool}} = \frac{5.58 + 2.52}{18}$$

$$\text{Step 4: } V_{\text{pool}} = \frac{8.10}{18}$$

$$\text{Step 5: } V_{\text{pool}} = .45$$

Now we can calculate the standard error of the mean difference $(V_{\overline{X}_1 - \overline{X}_2})$:

$$V_{\overline{X}_1 - \overline{X}_2} = \sqrt{(V_{pool})\left(\frac{1}{n_1} + \frac{1}{n_2}\right)}$$

Step 1: $V_{\overline{X}_1 - \overline{X}_2} = \sqrt{(.45)\left(\frac{1}{10} + \frac{1}{10}\right)}$

Step 2: $V_{\overline{X}_1 - \overline{X}_2} = \sqrt{(.45)(.10 + .10)}$

Step 3: $V_{\overline{X}_1 - \overline{X}_2} = \sqrt{(.45)(.20)}$

Step 4: $V_{\overline{X}_1 - \overline{X}_2} = \sqrt{.09}$

Step 5: $V_{\overline{X}_1 - \overline{X}_2} = .30$

And now we can calculate independent samples t test:

$$t = \frac{\overline{X}_1 - \overline{X}_2}{V_{\overline{X}_1 - \overline{X}_2}}$$

The mean for the experimental group $(\overline{X}_1)$ is 2.8 and the mean for the control group $(\overline{X}_2)$ is 2.5, so

Step 1: $t = \dfrac{2.8 - 2.5}{.30}$

Step 2: $t = \dfrac{.3}{.30}$

Step 3: $t = 1.00$

Using the t Tables

Since this is a one-tailed test, we will use the one-tailed t tables to determine if there is a significant relationship between the training and number of desks assembled. A portion of the t tables for a one-tailed test is reproduced in Table 12.2. Our formula for the degrees of freedom is

$$df = (n_1 - 1) + (n_2 - 1)$$

since we are comparing two groups. In a one-tailed test with 18 degrees of freedom and $\alpha = .05$, the critical t value is 1.734. This means that any t_{obt} that is greater than 1.734 is significant.

TABLE 12.2 One-Tailed *t* Table

df	$\alpha = .05$	$\alpha = .01$
15	1.753	2.602
16	1.746	2.583
17	1.740	2.567
18	1.734	2.552
19	1.729	2.539
20	1.725	2.528
21	1.721	2.518
22	1.717	2.508
23	1.714	2.500
24	1.711	2.492

Our t_{obt} of 1.00 is less than the t_{crit} of 1.734, so we retain the null hypothesis. We conclude that the training does not significantly increase the number of desks assembled.

THE POSSIBILITY OF SAMPLING ERROR

In each of the aforementioned tests we set our criteria at .05 because we can never be absolutely sure whether the independent variable is having an effect on the dependent variable because of the possibility of sampling error. For example, our two-tailed test demonstrated that the training has a positive effect. In other words, the people who are trained put together more desks that the people who did not receive the training.

However, we are dealing with people who were randomly selected and randomly assigned to the groups (experimental and control). We are 95% sure that the training has the desired effect, but there is the probability (.05) that the results we found were due to the fact that we somehow selected people who were already really good at assembling desks for the experimental group and the training actually had no (or very little effect). In other words, .05 represents the probability that our results were confounded by sampling error.

CHAPTER SUMMARY

The independent samples t test is conducted to compare two samples with each other. Samples are independent when being in one group means that you cannot be in the other. This is used to test the effect of an independent variable on an experimental group, comparing it to the control group that did not get the independent variable. If a significant effect is found, we measure the size of the effect by calculating the point biserial correlation coefficient.

FORMULAS FOR CHAPTER 12

The Estimated Population Variance: $v = \dfrac{\sum X^2 - \dfrac{(\sum X)^2}{N}}{N - 1}$

Pooled Variance: $V_{pool} = \dfrac{(n_1 - 1)(v_1) + (n_2 - 1)(v_2)}{(n_1 - 1) + (n_2 - 1)}$

Standard Error of the Mean Difference: $V_{\overline{X}_1 - \overline{X}_2} = \sqrt{(V_{pool})\left(\dfrac{1}{n_1} + \dfrac{1}{n_2}\right)}$

The Independent Samples t-Test: $t = \dfrac{\overline{X}_1 - \overline{X}_2}{V_{\overline{X}_1 - \overline{X}_2}}$

Degrees of Freedom: $df = (n_1 - 1) + (n_2 - 1)$

The Point Biserial Correlation Coefficient: $r_{pb}^2 = \dfrac{(t_{obt})^2}{(t_{obt})^2 + df}$

1. You are interested in whether male offenders are more or less dangerous than female offenders. You take a random sample of male and female offenders and compare their offense gravity score (OGS) to answer your question (higher scores indicate a more serious offense).

Males (Group 1)		Females (Group 2)	
6	6	6	5
4	10	4	5
7	10	4	6
10	9	3	9
12	10	4	4
12		7	

   a. Is this a one-tailed or a two-tailed test?
   b. What is the alternative hypothesis?
   c. What is the null hypothesis?
   d. With $\alpha = .05$, what is the t_{crit}?
   e. What is the t value (t_{obt})?
   f. Is there a difference between the seriousness of offense for the groups? If so, how much of the variability can be attributed to gender?

2. To test the idea that boys have less self-control than girls, a researcher uses the following scores on a test of self-control taken by both boys and girls:

Boys (Group 1)		Girls (Group 2)	
10	12	18	16
13	13	14	8
8	10	15	11
15	9	10	15
7	17	8	18

   a. Is this a one-tailed or a two-tailed test?
   b. What is the alternative hypothesis?
   c. What is the null hypothesis?
   d. With $\alpha = .05$, what is the t_{crit}?
   e. What is the t value (t_{obt})?
   f. Is there a difference between the levels of self-control of the groups? If so, how much of the variability can be attributed to gender?

3. To determine whether family environment effects juvenile delinquency, you compare the number of delinquent acts committed by juveniles from one-parent families to the number of delinquent acts committed by juveniles from two-parent families. You take a random sample of delinquent acts committed by children from each group and get the following results:

One Parent (Group 1)		Two Parents (Group 2)	
3	2	2	1
4	4	3	4
5	4	4	3
3	3	3	2
5	5	3	2
4	3	2	1
3	5	1	3

   a. Is this a one-tailed or a two-tailed test?
   b. What is the alternative hypothesis?
   c. What is the null hypothesis?
   d. With $\alpha = .05$, what is the t_{crit}?
   e. What is the t value (t_{obt})?
   f. Is there a difference between the two groups in terms of delinquency? If so, how much of the variability committed can be attributed to family environment?

4. A researcher wants to test the effect of type of media on knowledge of current political affairs. She randomly assigns one group of students to read only the internet and a second of group to read only daily newspapers for 60 days. She then administers a quiz about current political affairs to both groups and gets the following results:

Internet (Group 1)	Newspaper (Group 2)
6	5
5	6
5	4
4	4
5	4
6	6
5	5
4	5
5	4
5	4
5	4

a. Is this a one-tailed or a two-tailed test?
b. What is the alternative hypothesis?
c. What is the null hypothesis?
d. With $\alpha = .05$, what is the t_{crit}?
e. What is the t value (t_{obt})?
f. Is there a difference between knowledge of current political affairs? If so, how much of the variability can be attributed to type of media consumed?

5. You want to see if listening to music while studying has a detrimental effect on quiz grades. To test this, you randomly assign students to two groups. Group 1 spends 3 hours studying for a history quiz while listening to music, and group two spends 3 hours studying for the same history quiz while not listening to music. You collect the quizzes and get the following results:

Music (Group 1)	No Music (Group 2)
8	10
10	9
9	11
8	10
7	6
10	11
10	10

a. Is this a one-tailed or a two-tailed test?
b. What is the alternative hypothesis?
c. What is the null hypothesis?
d. With $\alpha = .05$, what is the t_{crit}?
e. What is the t value (t_{obt})?
f. Is there a difference between the quiz scores of the two groups? If so, how much of the variability can be attributed listening to music while studying?

THE RELATED SAMPLES *t* TEST

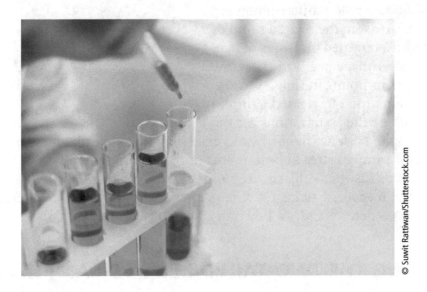

INTRODUCTION

As stated previously, one of the parts of a classical experiment is pre- and posttesting. This is when the participants in the experimental group are measured on the dependent variable before the experiment takes place, and then the same people are measured on the dependent variable after they have been exposed to the independent variable. If there is a significant difference between both sets of measures (pre and post), we can say that the difference is due to the effect of the independent variable. The test we use to determine statistically significant differences in this instance is the **related samples *t*-test**.

In the previous chapter, we learned how to compare two independent groups to each other, and in this chapter we will learn to compare two different scores to each other. The key difference is that here we will learn how to compare the same group at two different points in time[1].

---

[1] There is another related samples *t* test called the matched pair design, but since the calculations for this are the same as the related samples *t* test, and since the related samples *t* test is used more often, we will limit our discussion to the more commonly used procedure.

THE RELATED SAMPLES *t* TEST

Theoretically, in an experiment if the independent variable has no effect on the dependent variable then someone's score on a pretest should be the same on a posttest. For example (referring back to the example we used in Chapter 12), if someone can assemble two desks in 1 hour in a pretest, there is no reason to assume (absent an independent variable) that if asked to assemble desks for an hour again they should assemble a different number of desks. For this reason, the related samples *t* test is also called a repeated measures design. People in the experiment perform the same task at two different points in time.

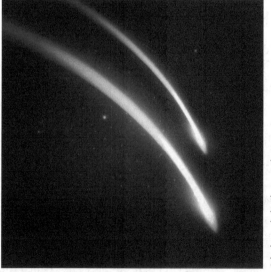

The assumptions for the related samples *t* test are the same as the assumptions for the independent samples *t* test, with the obvious exception that instead of having roughly equal samples sizes, the samples sizes at both points of the experiment will be exactly the same. The logic of this test is that subjects are randomly drawn from a population of interest to the researcher, measured twice on some dependent variable, the difference between the two measures is determined, and then the mean difference is calculated and analyzed.

CALCULATING THE TWO-TAILED RELATED SAMPLES *t* TEST

A teacher wants to test the effect of reading the newspaper on knowledge of current affairs. To do so, she randomly selects 20 students from her classes and gives them each a quiz on current affairs. She then instructs them to read only the newspaper for one week, and then she administers another quiz on current affairs at the end of the week.

Like always, we begin by setting up our testing, which begins with determining if we have a one- or a two-tailed test. Since we want to know if reading the newspaper "has an effect" on knowledge of current affairs, we have a two-tailed test. We are hypothesizing that scores between the first and second quizzes of current affairs (think, pre and posttests) will be different, but we are not saying in which direction (if they will be more knowledgeable or less).

Next we state our statistical hypotheses. We are comparing two sets of scores, but what is of interest to us is the difference between those scores, specifically, the average difference, which means we have to introduce a new symbol. Since $\bar{X}$ is the symbol for the mean (or average) of the raw scores, in our calculations we will use $\bar{D}$ to signify the average difference between scores. Since we are hypothesizing that reading the newspaper has an effect on quiz scores from the pretest to the posttest, we are saying that there will be a difference between them, so the average difference should be a number different than 0. If the scores were the same, or if there was no difference in the scores, the average difference would be 0. That means our alternative hypothesis is

$$H_A : \bar{D} \neq 0$$

which is another way of saying that there will be a difference between the scores (but, again, we are not saying which way).

Since the null hypothesis is also called the hypothesis of no difference, it states that the average difference between the pretest and posttests will be 0, so the null hypothesis is

$$H_0 : \bar{D} = 0$$

which states that the scores will not change from pre- to posttest and there will be no difference between the two measures. Reading the newspaper has no effect on knowledge of current affairs, so the scores from the pretest to the posttest will not change.

We calculate the mean difference much like we calculate the mean of raw scores, but in this case we add up all of the differences and divide by the sample size

$$\bar{D} = \frac{\sum D}{N}$$

In calculating our statistic, we need to arrange the data as follows:

Quiz 1 (Pretest)	Quiz 2 (Posttest)	D	D^2
5	5	0	0
6	6	0	0
6	7	1	1
5	7	2	4
7	7	0	0
4	6	2	4
4	5	1	1
5	6	1	1
8	7	−1	1

continued

Quiz 1 (Pretest)	Quiz 2 (Posttest)	D	D²
5	6	1	1
6	6	0	0
5	5	0	0
4	4	0	0
5	6	1	1
4	6	2	4
5	7	2	4
4	4	0	0
6	6	0	0
7	7	0	0
4	5	1	1
		$\Sigma D=13$	$\Sigma D^2=23$

Here we have the same students (or, test subjects) providing two scores. The convention with the related samples t test is to subtract the first score (pretest) from the second score (posttest) to get the values for D (difference) and D^2 (differences squared). This is so that when the mean difference is calculated, the direction of the change in scores (if there is one) can be determined. A positive value for the mean difference indicates that scores increased from pretest to posttest, and a negative value would mean that scores decreased from the pretest to the posttest. We square the differences and then sum them for a later calculation.

As you can see from the data, the mean difference equals 13/20, which is .65. Since this is a positive number, we can say that scores increased from the pretest to the posttest, but the question for us is did they increase because of the independent variable (reading a newspaper), or did they increase due to chance? We need to calculate the related samples t test to answer this.

Calculating the related samples t test is similar to calculating the one-sample t test. The formula for the related samples t test is

$$t = \frac{\bar{D}}{\upsilon_{\bar{D}}}$$

where we divide the mean difference by a statistic called the standard error of the mean difference.

With the one-sample t test, we started out computing the estimated population variance (υ) and then the estimated standard error of the mean ($S_{\bar{X}}$), so we could find the value for t (t_{obt}). Here, we will first calculate the estimated variance of the difference (υ_D), the formula for which is

$$\upsilon_D = \frac{\sum D^2 - \dfrac{\left(\sum D\right)^2}{N}}{N-1}$$

where N stands for the number of difference scores:

$$\text{Step 1: } \frac{23 - \frac{(13)^2}{20}}{19}$$

$$\text{Step 2: } \frac{23 - \frac{169}{20}}{19}$$

$$\text{Step 3: } \frac{23 - 8.45}{19}$$

$$\text{Step 4: } \frac{14.55}{19}$$

Answer: $v_D = .77$

Our next step is to calculate the standard error of the mean difference ($v_{\bar{D}}$), the formula for which is

$$v_{\bar{D}} = \sqrt{\frac{v_D}{N}}$$

To do this, we divide the estimated variance of the difference by the sample size, and then take the square root

$$\text{Step 1 : } \sqrt{\frac{.77}{20}}$$

$$\text{Step 2 : } \sqrt{.04}$$

Answer: $v_{\bar{D}} = .20$

We are now ready to calculate the related samples t test by dividing the mean difference by the standard error of the mean difference

$$t = \frac{\bar{D}}{v_{\bar{D}}}$$

$$t = \frac{.65}{.20}$$

Answer: $t_{obt} = 3.25$

Using the *t* Tables

Recall from Chapter 11 that there are two different *t* tables, depending on whether we have a one- or a two-tailed test. A portion of the *t* tables for a two-tailed test is reproduced in Table 13.1. We look up the critical *t* value by first finding the correct degrees of freedom in the column labeled *df*. There are two more columns for two different alpha levels (.05 and .01). We'll set our criterion at .05, so we will use that column to look up our critical *t* value. The formula for the degrees of freedom is

$$df = N - 1$$

where N is the number of difference scores. In this case, the degrees of freedom is equal to 19 (20 − 1). In a two-tailed test with 19 degrees of freedom and $\alpha = .05$, the critical *t* value is ±2.093 This means that any t_{obt} that is greater than 2.093 or less than -2.093 is statistically significant.

TABLE 13.1 Two-Tailed *t* Table

df	$\alpha = .05$	$\alpha = .01$
17	2.110	2.898
18	2.101	2.878
19	2.093	2.861
20	2.086	2.845
21	2.080	2.831
22	2.074	2.819

Our t_{obt} of 3.25 is greater than the t_{crit} of +2.093, so we reject the null hypothesis (reading the newspaper has no effect on knowledge of current affairs) and conclude that reading the newspaper has a statistically significant effect on knowledge of current affairs. In addition, since our t_{obt} is a positive number (hence, falls in the right-hand side of the *t* distribution), we can also conclude that reading the newspaper is likely to increase knowledge of current affairs.

EFFECT SIZE

When a significant relationship is found in a related samples *t* test, the last step is calculating the **effect size** of the independent variable (just like we did in Chapter 12). A significant result shows that the independent variable had an effect on the dependent variable. The question, then, is how much of an effect?

© Fourleaflover/Shutterstock.com

The larger the effect size, the greater the impact of the independent variable. Not all independent variables are created equal; even though there is an effect, this does not mean that the effect will be the same in all experiments. Remember that effect size is only calculated when there are significant results (when the null hypothesis is rejected). If the results are nonsignificant, then the independent variable did not have an effect on the dependent variable, so we do not bother looking for an effect size.

The Point Biserial Correlation Coefficient

As with the independent samples *t* test, we will be using the Point Biserial Correlation Coefficient with the related samples *t* test to measure how consistently changing conditions of the independent variable changes the conditions of the dependent variable. We measure this through the proportion of variance accounted for (as discussed in Chapter 8). The formula for the Point Biserial Correlation Coefficient is

$$r_{\text{pb}}^2 = \frac{(t_{\text{obt}})^2}{(t_{\text{obt}})^2 + df}$$

where we square the obtained t score and divide it by the squared obtained t score plus the degrees of freedom. For our example, the effect size of reading the newspaper between the pretest and posttest is

$$\text{Step 1: } r_{pb}^2 = \frac{(3.25)^2}{(3.25)^2 + 19}$$

$$\text{Step 2: } r_{pb}^2 = \frac{10.56}{10.56 + 19}$$

$$\text{Step 3: } r_{pb}^2 = \frac{10.56}{29.56}$$

$$\text{Step 4: } r_{pb}^2 = .36$$

This means that 36% of the variability in the quiz scores of knowledge of current affairs between the pretest and posttest is due to reading the newspaper. You could also say we improve our predictive ability of someone's knowledge of current affairs by 36% if we know that they read the newspaper. However, we always have to be cognizant of the possibility of sampling error. There is a 5% chance that the results we found are not attributable to the independent variable, but due to some other random factor.

CALCULATING THE ONE-TAILED t TEST

Another teacher hears of this experiment and they believe that going on the internet will increase knowledge of current affairs. She sets up a similar experiment where she takes a random sample of students and gives them a quiz on current affairs. Then she instructs them to go only to certain websites for one week, following which she administers another quiz on current affairs.

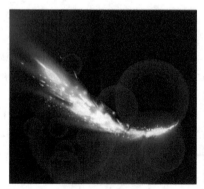

© Ron Dale/Shutterstock.com

This is a one-tailed test (increase knowledge) because here the teacher is predicting the direction of the relationship. This makes the alternative hypothesis

$$H_A : \bar{D} > 0$$

since she is predicting that the scores on the posttest will be greater than the scores on the pretest. Therefore, subtracting the pretest scores from the posttest scores will result in a number greater than 0. Conversely, the null hypothesis is

$$H_0 : \bar{D} \leq 0$$

because the hypothesis of no difference states that the independent variable (in this case, the internet) will have no effect on the dependent variable (knowledge of current affairs), and so the posttest scores will be the same as, or less than, the pretest scores, making the average differences 0 or a negative number (less than 0). The data are collected, and they look like this

Quiz 1 (Pretest)	Quiz 2 (Posttest)	D	D²
6	6	0	0
5	5	0	0
7	8	1	1
6	6	0	0
8	8	0	0
8	9	1	1
6	6	0	0
5	5	0	0
4	4	0	0
4	4	0	0
3	3	0	0
7	7	0	0
6	6	0	0
5	5	0	0
5	5	0	0
		$\Sigma D=2$	$\Sigma D^2=2$

The mean difference ($\bar{D}$) is .13 (2/15). This is obviously greater than 0, but is it enough to say that the internet increases knowledge of current affairs? Only the related samples *t* test can answer this for us.

We first calculate the estimated variance of the difference (v_D)

$$v_D = \frac{\sum D^2 \dfrac{\left(\sum D\right)^2}{N}}{N-1}$$

$$\text{Step 1}: \frac{2-\dfrac{(2)^2}{15}}{14}$$

$$\text{Step 2}: \frac{2-\dfrac{4}{15}}{14}$$

$$\text{Step 3}: \frac{2-.27}{14}$$

$$\text{Step 4}: \frac{1.73}{14}$$

$$\text{Answer}: v_D = .12$$

Our next step is to calculate the standard error of the mean difference ($v_{\bar{D}}$)

$$v_{\bar{D}} = \sqrt{\frac{v_D}{N}}$$

$$\text{Step 1}: \sqrt{\frac{.12}{15}}$$

$$\text{Step 2}: \sqrt{.008}$$

$$\text{Answer}: v_{\bar{D}} = .09$$

The last step is to calculate the related samples t test by dividing the mean difference by the standard error of the mean difference

$$t = \frac{\bar{D}}{v_{\bar{D}}}$$

$$t = \frac{.13}{.09}$$

$$\text{Answer}: t_{\text{obt}} = 1.44$$

To determine if this is a statistically significant number, we go back to the t tables. Since this is a one-tailed test, we have to consult the one-tailed t table, a portion of which is reproduced in table 13.2

TABLE 13.2 One-Tailed *t* Table

df	$\alpha = .05$	$\alpha = .01$
12	1.782	2.681
13	1.771	2.650
14	1.761	2.624
15	1.753	2.602
16	1.746	2.583
17	1.740	2.567

With 14 degrees of freedom ($N - 1$) and $\alpha = .05$, our critical t value is 1.761. Our calculated t value (t_{obt}) is less than this, so we retain the null hypothesis. The internet has no effect on knowledge of current affairs.

CHAPTER SUMMARY

We use the related samples t test when we want to compare the same test subject at different points in time, specifically before and after the administration of some independent variable. The related samples t test is a parametric inferential procedure, which when used in conjunction with the independent samples t test, gives us a good idea of the effectiveness of an independent variable in an experimental setting.

FORMULAS FOR CHAPTER 13

The Estimated Variance of the Difference: $v_D = \dfrac{\sum D^2 - \dfrac{\left(\sum D\right)^2}{N}}{N-1}$

Standard Error of the Mean Difference: $v_{\bar{D}} = \sqrt{\dfrac{v_D}{N}}$

The Related Samples t Test: $t = \dfrac{\bar{D}}{v_{\bar{D}}}$

Degrees of Freedom: $df = (N-1)$

The Point Biserial Correlation Coefficient: $r_{pb}^2 = \dfrac{(t_{obt})^2}{(t_{obt})^2 + df}$

1. A researcher wants to study whether watching a violent television show will have an effect the behavior of a group of children. She watches a group of children play for an hour, recording the number of violent acts they perform. Then she has them watch a violent television show, then play for another hour, again recording the number of violent acts. The number of violent acts committed before and after the violent show are recorded as follows:

Before Show	After Show
4	3
3	3
4	5
4	6
7	8
3	5
0	2
0	1
5	6
4	3

   a. Is this a one-tailed or a two-tailed test?
   b. What is the alternative hypothesis?
   c. What is the null hypothesis?
   d. With $\alpha = .05$, what is the t_{crit}?
   e. What is the t value (t_{obt})?
   f. Is there a difference between number of violent acts committed before and after the violent show? If so, how much of the variability can be attributed to watching the show?

2. A researcher develops a pill that will assist with weight loss, so she samples a group of people to test the pill. Participants are weighed before they take the pill, and then one week later. The results are as follows:

Weight Before	Weight After
157	143
190	175
140	142
165	150
112	110
149	141
173	157
128	132

a. Is this a one-tailed or a two-tailed test?
b. What is the alternative hypothesis?
c. What is the null hypothesis?
d. With $\alpha = .05$, what is the t_{crit}?
e. What is the t value (t_{obt})?
f. Is there a difference between the bodyweights of participants before and after taking the weight loss pill? If so, how much of the variability in their weights can be attributed to the pill?

3. Researchers want to investigate whether implementing a neighborhood watch program has an effect on the reported acts of vandalism in a neighborhood. They randomly sample 15 neighborhoods that do not have a neighborhood watch program, implement one, and record the number of acts of vandalism in those neighborhoods for one year before implementing the program and one year later. They obtain the following results:

Before	After
7	5
6	6
5	5
8	6
7	5
7	3
8	5
10	10
11	12
10	9
8	10
7	4
6	3
6	3
7	4

a. Is this a one-tailed or a two-tailed test?
b. What is the alternative hypothesis?
c. What is the null hypothesis?
d. With $\alpha = .05$, what is the t_{crit}?
e. What is the t value (t_{obt})?
f. Is there a difference between the two measures in terms of acts of vandalism? If so, how much of the variability can be attributed to the neighborhood watch program?

4. Doctors develop a sleeping pill to help people sleep more. To test the pill, they take a sample of volunteers and have them take an average of the number of hours they sleep per night for one week. They then administer the pill and have the participants average the number of hours of sleep they get per night for the following week. The results are as follows:

Before Pill	After Pill
6	6
4	7
4	7
6	7
6	8
9	8
6	6
7	10
10	11
9	8
8	9
5	7

a. Is this a one-tailed or a two-tailed test?
b. What is the alternative hypothesis?
c. What is the null hypothesis?
d. With $\alpha = .05$, what is the t_{crit}?
e. What is the t value (t_{obt})?
f. Is there a difference between amount of sleep per night between the pretest and the posttest? If so, how much of the variability can be attributed to the sleeping pill?

THE ONE-WAY
ANALYSIS OF VARIANCE

© christitzeimaging.com/Shutterstock.com

INTRODUCTION

In this chapter we are going to discuss another parametric inferential procedure called the **analysis of variance,** or ANOVA. The ANOVA is used when we have an experiment with more than one condition of the independent variable, which makes it one of the most used inferential procedures in Behavioral Sciences research. There are different versions of the ANOVA because it is used with different experimental designs.

In the last chapter we examined the independent samples *t*-test, which we use to determine if the difference between two groups is due to an independent variable (a training on desk assembly), or if any differences are attributable to random chance. Sometimes, though, researchers want to compare more than two groups. Let us say the company that makes desks wants to compare three groups: one group that gets a one-hour training (group 1); another that completes a 2-hour training (group 2); and a third that gets a three-hour training (group 3).

One way to see if there is a significant difference between these groups is to conduct three independent samples *t* tests. We could compare groups 1 and 2, then groups 1 and 3, and then groups 2 and 3. However, the reason we do not do this is because conducting multiple *t* tests separately increases the probability of committing a Type I Error. This is because of something called the **experiment-wise error rate,** which is the probability of making a Type I Error anywhere among the comparisons of different groups in an experiment. In other words, we could make a Type I Error when comparing groups 1 and 2, when comparing groups 1 and 3, and when comparing groups 2 and 3. The probability of making at least one error is greater than .05 (assuming the researcher sets their alpha at .05), or one in 20. If we run three independent samples *t* tests, our collective error increases to 3 in 20 (or .15). So what we do rather than conduct multiple independent samples *t* tests is to conduct an ANOVA, which keeps the experiment-wise error rate at .05 because we are measuring the differences between all of the groups at once.

While we can use the ANOVA to compare more than two groups, we are still examining the effect of one independent variable (with more than two conditions) on a dependent variable. Therefore, we will be conducting a **one-way analysis of variance,** since there is one independent variable. The logic is the same as an independent samples *t* test, we are just extending it to more than two groups.

THE ONE-WAY ANOVA

The one-way analysis of variance is a parametric inferential procedure that we use when we are interested in evaluating whether more than two sample means differ due to actual differences between them or merely by chance. In other words, we are determining whether we have significant differences in an experiment with more than two conditions (which we will call treatments). Like all of the other statistics we have discussed in this book, there are assumptions that must be met before we can use the one-way ANOVA:

© LDDesign/Shutterstock.com

1. The dependent variable is measured at the interval/ratio level.
2. The dependent variable is normally distributed in the population.
3. The populations from which the samples are taken have homogeneous variances (the variances of the populations being represented are equal).
4. The experiment has only one independent variable and all conditions contain independent samples.
5. Random sampling.
6. Group assignment is random.

CALCULATING THE ONE-WAY ANOVA

We are going to stick with our desk manufacturer example that we used in last chapter. Now instead of testing whether a training on putting desks together has an effect on the number desks assembled, let us say that the manufacturer wants to determine the ideal length of such a training. The manufacturer finds volunteers to be a part of this experiment, and the volunteers are randomly assigned to three different groups: Group 1 will be trained for one hour; Group 2 will

receive 2 hours of training; and Group 3 will be trained for 3 hours. The purpose is to see if the length of the training (independent variable) has an effect on the number of desks the volunteers can assemble (dependent variable). Fifteen people volunteer, so each group has five participants. Keep in mind that there is one independent variable (the training) with three conditions (1 hour, 2 hours, and 3 hours). To determine whether there is a significant difference between the lengths of the trainings, we will conduct a one-way ANOVA.

Our statistic is called analysis of variance because that is exactly what it does—it analyzes variance. With the ANOVA, there are two types of variance: **within-treatment variance** and **between-treatments variance**. Within-treatment variance refers to the variance that occurs within a particular sample. There are two sources of this variability, individual differences and experimental error.

Let us say that in a group that is trained on putting desks together, different people assemble different numbers of desks. This variation can be due to individual differences, things like whether someone works with their hands for a living, previous experience with desk assembly, etc. However, any differences could also be due to experimental error, which is anything in the experiment that could account for differences within the sample, things like poor lighting, background noises, etc. Both types of variance are random because they are unintentional.

Between treatments variance refers to the differences between the three samples. If each group has a different sample mean (each group assembles, on average, a different number of desks), it could be due to the two types of variance previously discussed. Or, the difference in sample means (mean number of desks assembled in each treatment condition) may be due to the different levels of the independent variable or, in other words, a treatment effect.

If there are no treatment effects, the between-treatments variance and within treatment variance would be roughly equal. However, if the treatment is effective then the within treatment variance would be smaller than the between-treatments variance. This is due to the influence of the treatment effect.

Now, to test whether the different levels of the training have any effect of the number of desks assembled, we need to set up our procedures, starting with the null and alternative hypotheses. Remember, the null hypothesis is also called the hypothesis of no difference, so in this case it would hypothesize that the training has no effect on the number of desks assembled. If this were the case, we would not expect a difference in the number of desks assembled by each group, so the null hypothesis is

$H_0 : \bar{X}_1 = \bar{X}_2 = \bar{X}_3 = \bar{X}_k$ (the subscript k means that there will be as many sample means as there are groups)

The alternative hypothesis asserts that there will be statistically significant differences between some of the sample means, but we are not sure which ones. For example, there could be a difference in the mean number of desks assembled between groups 1 and 2, but there could be no difference between groups 1 and 3. Therefore we use a generic alternative hypothesis, which is

$H_A :$ Some $\bar{X}$s are not equal

Notations for the One-Way ANOVA

Calculating the one-way ANOVA entails working through a number of steps. Here are the notations that we will need:

- **Sums of Squares (SS)** is the sum of the squared deviations of scores from the mean. To do this we need to calculate three SS values: SS_{tot}, SS_{wn}, and SS_{bn}. SS_{tot} is the total sum of squared deviations, or the total variance for all of the groups. SS_{wn} is the sum of squared deviations within each treatment group, or the internal group variation. Finally, SS_{bn} is the sum of squared deviations between each treatment group, or the variation of each group mean from the overall mean for the study. The sums of squares represent the effect of an independent variable and the overall variability within the groups when they are compared with each other.
- ΣX_{tot} is the total of all of the scores in a study. To get this value, we will add the sum of all of the X scores for each treatment group.
- ΣX^2_{tot} is the total of all the squared scores in the experiment. We calculate this by adding the sum of the squared scores for each group.
- N is the total number of all scores in the study.
- n is the number of scores in a sample group. Each group is designated by different subscripts.
- k is the number of groups in the study.

Our data meet the assumptions for a one-way ANOVA, we have our statistical hypotheses, so let us set alpha at .05 and conduct the test. The fifteen volunteers are randomly assigned to three different groups, trained for different amounts of time, and then assemble the following number of desks:

Group 1 (1 hour)		Group 2 (2 hours)		Group 3 (3 hours)	
X	X^2	X	X^2	X	X^2
3	9	8	64	3	9
3	9	12	144	7	49
4	16	9	81	10	100
6	36	7	49	6	36
4	16	9	81	9	81

$$\Sigma X = 20$$
$$\Sigma X^2 = 86$$
$$n = 5$$
$$\overline{X}_1 = 4$$

$$\Sigma X = 45$$
$$\Sigma X^2 = 419$$
$$n = 5$$
$$\overline{X}_2 = 9$$

$$\Sigma X = 35$$
$$\Sigma X^2 = 275$$
$$n = 5$$
$$\overline{X}_3 = 7$$

$$\Sigma X_{tot} = 100$$
$$\Sigma X^2_{tot} = 780$$
$$N = 15$$
$$k = 3$$

As you can see the sample means for each treatment condition are different, which would indicate that we should reject the null hypothesis. However, we need to establish if these differences are statistically significant, meaning that the differences are attributable to the different levels of the independent variable, or if they are due to random chance.

The first calculation in the one-way ANOVA is the three sum of squares values. First, we will calculate the SS_{tot}, which is the sum of squares for all of the scores, the formula for which is

$$SS_{tot} = \sum X^2_{tot} - \frac{\left(\sum X_{tot}\right)^2}{N}$$

so we insert the values as follows:

$$SS_{tot} = 780 - \frac{(100)^2}{15}$$

$$SS_{tot} = 780 - \frac{10,000}{15}$$

$$SS_{tot} = 780 - 666.67$$

$$SS_{tot} = 113.33$$

Next we calculate the SS_{wn}, the formula for which is

$$SS_{wn} = \sum \left[\sum X^2 - \frac{(\sum X)^2}{n} \right]$$

We have to calculate the sum of squares within each treatment group, then sum each of the values

$$SS_{wn} = \left[86 - \frac{(20)^2}{5} \right] + \left[419 - \frac{(45)^2}{5} \right] + \left[275 - \frac{(35)^2}{5} \right]$$

$$SS_{wn} = \left[86 - \frac{400}{5} \right] + \left[419 - \frac{2025}{5} \right] + \left[275 - \frac{1225}{5} \right]$$

$$SS_{wn} = [86 - 80] + [419 - 405] + [275 - 245]$$

$$SS_{wn} = 6 + 14 + 30$$

$$SS_{wn} = 50$$

Next we will calculate the SS_{bn}, the formula for which is

$$SS_{bn} = \sum \left[\frac{(\sum X)^2}{n} \right] - \frac{(\sum X_{tot})^2}{N}$$

Here, we perform the operation in the brackets for each group before subtracting the expression at the end. Inserting the numbers looks like this

$$SS_{bn} = \left[\frac{20^2}{5} + \frac{45^2}{5} + \frac{35^2}{5} \right] - \frac{100^2}{15}$$

$$SS_{bn} = \left[\frac{400}{5} + \frac{2,025}{5} + \frac{1,225}{5} \right] - \frac{10,000}{15}$$

$$SS_{bn} = [80 + 405 + 245] - 666.67$$

$$SS_{bn} = 730 - 666.67$$

$$SS_{bn} = 63.33$$

A way to quality control yourself here is that the sum of squares within variability and sum of squares between variability should equal the sum of squares total variability. Here, $SS_{wn} = 50$ plus $SS_{bn} = 63.33$ equals $SS_{tot} = 113.33$. We are on the right track.

The Degrees of Freedom

We need to determine the degrees of freedom for each of the three sums of squares that we calculated, each of which have a different formula:

- $df_{tot} = N - 1$, so for our problem $df_{tot} = 15 - 1 = 14$
- $df_{wn} = N - k$, so for our problem $df_{wn} = 15 - 3 = 12$
- $df_{bn} = k - 1$, so for our problem $df_{bn} = 3 - 1 = 2$

Mean Squares (MS)

In the one-way ANOVA, the **mean squares** is calculated by dividing each sum of squares (SS) by its degrees of freedom (df). We calculate the between mean square and within mean square for our data as follows

$$MS_{bn} = \frac{SS_{bn}}{df_{bn}} = \frac{63.33}{2} = 31.67$$

$$MS_{wn} = \frac{SS_{wn}}{df_{wn}} = \frac{50}{12} = 4.17$$

In our example, there are three groups being compared, so they make up the sources of between-groups variation. The within-group variation is the product of all of the cases in the data (or the sample size, which in our example is 15), so there are more cases contributing to the total variation. Our final result is a standardized comparison of the two.

The *F*-Statistic

The final calculation for the one-way ANOVA is the *F*-statistic. The *F*-statistic is a measure of the treatment variance divided by the error variance

$$F = \frac{\text{Treatment variance}}{\text{Error variance}}$$

Earlier we said that if there are no treatment effects, the between treatment variance and within treatment variance would be roughly equal. If the null hypothesis is true, then the value we get for the treatment variance will be the same number as the number we get for the error variance. In other words, if the value for the numerator is the same as the value as the denominator, the value for *F* should be 1 if the null hypothesis is correct.

We will reformulate the above equation to make it more user-friendly for our calculations. Since MS_{bn} is a measure of treatment variance, and MS_{wn} is a measure of error variance, our formula for the *F* statistic is

$$F = \frac{MS_{bn}}{MS_{wn}}$$

so for our data the value for F is

$$F = \frac{31.67}{4.17} = 7.59$$

To simplify this, we use an ANOVA summary table. For our problem, it looks like Table 14.1.

TABLE 14.1 Summary Table of One-Way ANOVA

Source	Sum of Squares	df	Mean Square	F
Between	63.33	2	31.67	7.59
Within	50.00	12	4.17	
Total	113.33	14		

To determine whether our calculated F statistic is statistically significant and if we should reject the null hypothesis, we need to compare it to a critical value of F. To do this, use the F Tables which we will find in Appendix A, a portion of which is reproduced in Table 14.2. It is important to note that since F values are calculated from squared scores, there will never be a negative F statistic.

TABLE 14.2 The F-Tables

df Within Groups	α	df Between Groups				
		1	2	3	4	5
8	.05	5.32	4.46	4.07	3.84	3.69
	.01	11.26	8.65	7.59	7.01	6.63
9	.05	5.12	4.26	3.86	3.63	3.48
	.01	10.56	8.02	6.99	6.42	6.06
10	.05	4.96	4.10	3.71	3.48	3.33
	.01	10.04	7.56	6.55	5.99	5.64
11	.05	4.84	3.98	3.59	3.36	3.20
	.01	9.65	7.20	6.22	5.67	5.32
12	.05	4.75	3.88	3.49	3.26	3.11
	.01	9.33	6.93	5.95	5.41	5.06
13	.05	4.67	3.80	3.41	3.18	3.02
	.01	9.07	6.70	5.74	5.20	4.86

As you can see, there are two areas for degrees of freedom. This is because when establishing the critical value we need to consider both the within-group degrees of freedom (df_{wn}) and the between-groups degrees of freedom (df_{bn}). The df_{wn} can be found on the left side of the table, and the df_{bn} is found at the top. We match the df_{wn} for our problem, which is 12, with the df_{bn} for our problem, which is 2, and we see that for an alpha of .05 the critical value for F is 3.88.

As in previous chapters when we had to establish statistical significance using a statistical table, if the calculated value, or F_{obt} in this case, is greater than the critical value, or F_{crit}, we reject the null hypothesis. If the calculated value is less than the critical value, we retain the null hypothesis. The idea is that differences that small would have occurred by chance and are not due to the influence of the independent variable. In this case, our obtained value of 7.59 is greater than the critical value of 3.88, so we reject the null hypothesis that there is no difference between the sample means.

A significant result does not tell us which groups differ. Our significant ANOVA indicates that somewhere among the groups being compared there is at least one significant difference. If we have a significant value for F_{obt}, our next step is to determine which group comparisons are significantly different. To do this, we conduct a post-hoc test.

Post Hoc Comparisons

Since we rejected the null hypothesis, we now look at the three sample means.

Group 1 (1 hour)	Group 2 (2 hours)	Group 3 (3 hours)
$\bar{X}_1 = 4$	$\bar{X}_2 = 9$	$\bar{X}_3 = 7$

These means are all different from each other, but what we need to do now is determine where the statistically significant differences are. To do this, we will perform a post-hoc analysis. As mentioned in previous chapters, we only perform post hoc analyses when we reject the null hypothesis. If we retained the null (there are no differences in the treatment levels), we would not perform a post-hoc analysis. Though there is more than one post hoc test for the one-way ANOVA, we will conduct the Tukey HSD multiple comparison test since it is used when your sample sizes are all equal.

Tukey's HSD Multiple Comparison Test

The Tukey HSD multiple comparison test (HSD) computes the minimum difference between two means that is required for us to say that they are significantly different. The first step in computing the HSD is to find the value of q_k in the table "Values of Studentized Range Statistic" in Appendix A of this book, a portion of which is reproduced in Table 14.3.

TABLE 14.3 Values of Studentized Range Statistic

df Within Groups	α	\multicolumn{5}{c}{k = Number of Means Being Compared}				
		2	3	4	5	6
8	.05	3.26	4.04	4.53	4.89	5.17
	.01	4.74	5.63	6.20	6.63	6.96
9	.05	3.20	3.95	4.42	4.76	5.02
	.01	4.60	5.43	5.96	6.35	6.66
10	.05	3.15	3.88	4.33	4.65	4.91
	.01	4.48	5.27	5.77	6.14	6.43
11	.05	3.11	3.82	4.26	4.57	4.82
	.01	4.39	5.14	5.62	5.97	6.25
12	.05	3.08	3.77	4.20	4.51	4.75
	.01	4.32	5.04	5.50	5.84	6.10
13	.05	3.06	3.73	4.15	4.45	4.69
	.01	4.26	4.96	5.40	5.73	5.98

We find the correct value by using the number of groups on our experiment (k) and the within-groups degrees of freedom. For our example $k = 3$, $df_{wn} = 12$, and $\alpha = .05$, so $q_k = 3.77$.

The next step is to compute HSD, the formula for which is

$$HSD = (q_k)\left(\sqrt{\frac{MS_{wn}}{n}}\right)$$

Inserting the numbers for our problem looks like this

$$HSD = (3.77)\left(\sqrt{\frac{4.17}{5}}\right)$$

$$HSD = (3.77)\left(\sqrt{.834}\right)$$

$$HSD = (3.77)(.91)$$

$$HSD = 3.43$$

The next step is to use HSD to determine the differences between each pair of means. We subtract each mean from every other mean (it does not matter if the number is negative or positive). For our example, it is

$$\bar{X}_1 = 4 - \bar{X}_2 = 9 = -5 \text{ (absolute value} = 5)$$

$$\bar{X}_1 = 4 - \bar{X}_3 = 7 = -3 \text{ (absolute value} = 3)$$

$$\bar{X}_2 = 9 - \bar{X}_3 = 7 = 2 \text{ (absolute value} = 2)$$

In the last step, we compare the absolute value of the differences between each mean to the value for HSD, which for this example is 3.43. Any mean difference greater than the HSD value is statistically significant. This means, looking at the numbers above, there is a significant difference between group 1 (one hour) and group 2 (two hours) because the absolute difference between the group means (5) is greater than our HSD value (3.77). This is the only significant difference, since the absolute values of the other two differences (groups 1 and 3, and groups 3 and 2) are less than 3.77. For the desk manufacturer, this means that a training of 2 hours makes more sense than 1 or 3 hours.

Determining Effect Size

The final part of conducting a one-way ANOVA is to conduct one more post hoc test, this time measuring effect size. Again, we only conduct post hoc tests if we reject the null hypothesis. In this case, we have established that the length of training has an effect on the number of desks assembled, so now we want to know how much of an effect. To determine effect size, we will calculate **eta squared**, the formula for which is

$$\acute{\eta}^2 = \frac{SS_{bn}}{SS_{tot}}$$

Remember, the sum of squares (SS) is a measure of variability. SS_{tot} measures the total variability in an experiment and SS_{bn} measures the variability between groups, which is a result of different treatment levels. What eta squared tells us, then, is the proportion of the total variability (SS_{tot}) that is accounted for by the independent variable (SS_{bn}). For our example, it is

$$\acute{\eta}^2 = \frac{63.33}{113.33}$$

$$\acute{\eta}^2 = .56$$

What this means is that 56% of the variability in scores (in this case, number of desks assembled) can be explained by the treatment (length of training). In other words, the manufacturer can increase the number of desks assembled by 56% if their employees receive 2 hours of training on desk assembly. Two hours is significantly better than 1 hour of training, and training for 3 hours will not have any appreciable effect.

CHAPTER SUMMARY

The one-way ANOVA is a very useful test of significance. It is conducted when we want to determine the effect of more than one condition of one independent variable on a dependent variable. The assumptions of the one-way ANOVA are similar to the assumptions of the independent samples t-test. We calculate the F statistic and compare it to a critical value to determine significant differences between groups. If there are significant differences, we then conduct two post hoc tests, one to determine where the differences are, and another to determine the effect size of the independent variable.

FORMULAS FOR CHAPTER 14

Sum of Squares Total $= SS_{tot} = \sum X_{tot}^2 - \dfrac{\left(\sum X_{tot}\right)^2}{N}$

Sum of Squares Within Groups $= SS_{wn} = \sum\left[\sum X^2 - \dfrac{(\sum X)^2}{n}\right]$

Sum of Squares Between Groups $= SS_{bn} = \sum\left[\dfrac{(\sum X)^2}{n}\right] - \dfrac{(\sum X_{tot})^2}{N}$

Degrees of Freedom Total $= df_{tot} = N - 1$

Degrees of Freedom Within Groups $= df_{wn} = N - k$

Degrees of Freedom Between Groups $= df_{bn} = k-1$

Mean Squares Between Groups $= MS_{bn} = \dfrac{SS_{bn}}{df_{bn}}$

Mean Squares Within Groups $= MS_{wn} = \dfrac{SS_{wn}}{df_{wn}}$

F Statistic $= F = \dfrac{MS_{bn}}{MS_{wn}}$

Tukey HSD $= HSD = (q_k)\left(\sqrt{\dfrac{MS_{wn}}{n}}\right)$

Eta Squared $= \acute{\eta}^2 = \dfrac{SS_{bn}}{SS_{tot}}$

1. A researcher investigated the number of reports of police officer misconduct as a function of officer-reported on-the-job stress and got the following results

Minimal Stress	Moderate Stress	Severe Stress
2	4	6
1	3	5
4	2	7
1	3	5

   a. What is the null hypothesis?
   b. What is the alternative hypothesis?
   c. Compute F_{obt} and complete the ANOVA summary table.
   d. With $\alpha = .05$, what is the critical value of F (F_{crit})?
   e. Is there a relationship between on-the-job stress and number of reports of police misconduct?
   f. If your results are significant, perform the appropriate post hoc test.
   g. If your results are significant, compute and interpret the effect size.

2. A researcher investigated the relationship between the number of convictions for white-collar crime violations and offense type and got the following results

Embezzlement	Bribery	Anti-Trust
9	8	10
10	9	9
12	12	11
11	12	8
9	8	9
11	10	9

   a. What is the null hypothesis?
   b. What is the alternative hypothesis?
   c. Compute F_{obt} and complete the ANOVA summary table.
   d. With $\alpha = .05$, what is the critical value of F (F_{crit})?
   e. What do you conclude about the relationship between convictions for white-collar crime and offense type?
   f. If your results are significant, perform the appropriate post hoc test.
   g. If your results are significant, compute and interpret the effect size.

199

3. You investigate the relationship between incidents of self-reported depression by inmates and the type of facility they are in, so you collect the following data

Minimum Security	Medium Security	Maximum Security
3	4	5
3	4	4
3	4	4
2	3	3
1	5	5

a. What is the null hypothesis?
b. What is the alternative hypothesis?
c. Compute F_{obt} and complete the ANOVA summary table.
d. With $\alpha = .05$, what is the critical value of F (F_{crit})?
e. What do you conclude about the relationship between self-reported depression and type of facility?
f. If your results are significant, perform the appropriate post hoc test.
g. If your results are significant, compute and interpret the effect size.

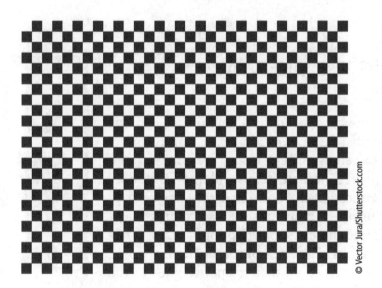

© Vector Jura/Shutterstock.com

INTRODUCTION

In the previous five chapters we discussed parametric inferential procedures (*z* test, one-sample *t* test, the independent samples *t* test, the related samples *t* test, and the one-way ANOVA) that are used when our data meet certain assumptions. We use nonparametric inferential procedures when our data do not meet the assumptions of parametric procedures, specifically when the dependent variable is skewed interval or ratio level data, or when it is measured at the nominal or ordinal level. Nonparametric procedures are also called distribution-free statistics because you do not need a normal distribution to use them. In this chapter, we are going to explore the chi-square (pronounced kai-square) test, which is a nonparametric inferential procedure, where the dependent variable is measured at the nominal level.

If you recall from Chapter 2, nominal-level variables are also called categorical variables because different values represent different categories. For example, college major is a nominal-level variable because it does not represent the amount of something but only the categories (criminal justice, behavioral sciences, nursing, etc.) to which people belong. There are two different chi-square tests, depending on whether we are examining one variable or two.

There are two assumptions for the chi-square test, and they are as follows:

1. Category membership is independent, which means that if you are in one category you cannot be in another.
2. The expected frequencies (we will get to that shortly) for all categories must be at least five.

CALCUATING THE ONE-WAY CHI-SQUARE

© Zoart Studio/Shutterstock.com

We perform the one-way chi-square when the data consists of one variable with at least two categories. For example, we ask a sample of people in a city if they are going to vote for a certain political candidate (let us call him Candidate A). In asking this question, we provide three response options: yes, no, or unsure (adding unsure makes our choices exhaustive). Think of these response options as categories to which people belong. For example, if a person is going to vote for Candidate A, then their response goes in the "yes" category, and so on. We sample 1,000 people and get the following responses:

Will You Vote for Candidate A?

Yes	No	Unsure
380	319	301

Our job now is to determine whether the differences we see in the categories are statistically significant, meaning, do these differences reflect actual voting preferences or are they due to random chance. Since our variable is measured at the nominal level, and there is only one variable (with three attributes), we compute the one-way chi-square.

There are statistical hypotheses with the chi-square test, but they look very different than the statistical hypotheses used with parametric procedures. Since our data are measured at the nominal level, we cannot compute a mean, so we will not see any sample means or population means. Instead we have to write out the hypotheses. The alternative hypothesis states that there are differences in the categories

H_A: There are differences in voter preferences.

The alternative hypothesis for the one-way chi-square always starts with the words "There are differences" and ends with the variable under consideration. If we were measuring attitude towards the legalization of marijuana, then our alternative hypothesis would be "There are differences in attitudes toward marijuana legalization."

Remember that the null hypothesis is also called the hypothesis of no difference. Therefore, the null hypothesis for this problem is

H_0: There are no differences in voter preferences.

The null hypothesis for the one-way chi-square always starts with the words "There are no differences" and ends with the variable under consideration. To use the example from above, the null hypothesis would state that there are no differences in attitudes toward marijuana legalization.

There is only one chi-square formula

$$X^2 = \sum \left(\frac{(f_o - f_e)^2}{f_e} \right)$$

where f_o means **observed frequencies** and f_e means **expected frequencies**. In the numerator, we subtract the expected frequencies from the observed frequencies for each cell (or category—I will use the words "cell" and "category" interchangeably), square that number, and then divide it by the expected frequencies. We then add each of these numbers together (hence the Greek Σ meaning sum outside of the parentheses) to get our chi-square statistic.

The observed frequencies are the numbers in each category from the sample. We asked 1,000 people if they were going to vote for Candidate A, and 380 people said yes, 319 said no, and 301 said they were unsure. The expected frequencies are the frequencies that we would expect to find if the null hypothesis is correct. The expected frequencies have to be calculated, and the formula for expected frequencies in a one-way chi-square is

$$f_e = N/K$$

where N is the sample size and K is the number of categories. So the expected frequencies for the categories of voter preference are

$$f_e = 1{,}000/3$$
$$f_e = 333.33$$

As you can see, the expected frequencies will not always be a whole number. Since the null hypothesis says that there are no differences in voter preference, in the one-way chi-square the expected frequencies will always be the same the number (this will not be the case with the two-way chi-square). So the observed frequencies and expected frequencies (in bold) for voter preference look like this

Will You Vote for Candidate A?

Yes		No		Unsure	
380	**333.33**	319	**333.33**	301	**333.33**

Obviously there are differences in each cell, but the questions for us are (1) are these differences significant enough that we would say they are not happening by chance, and (2) is this how the entire population would vote? Keep in mind, the chi-square is an inferential procedure, so we use it to infer from our sample to the population.

Now that we have the expected and observed frequencies, we can calculate the chi-square statistic. We do this going cell by cell so for the Yes category the chi-square is

$$X^2 = \frac{(380 - 333.33)^2}{333.33} = \frac{(46.67)^2}{333.33} = \frac{2,178.09}{333.33} = 6.53$$

Next, we compute the chi-square for the No category

$$X^2 = \frac{(319 - 333.33)^2}{333.33} = \frac{(-14.33)^2}{333.33} = \frac{205.35}{333.33} = .62$$

And finally the chi-square for the Unsure category

$$X^2 = \frac{(301 - 333.33)^2}{333.33} = \frac{(-32.33)^2}{333.33} = \frac{1,045.23}{333.33} = 3.14$$

The last step is to add (remember, Σ) the chi-square for each cell to give us the overall chi-square statistic:

$$X^2 = 6.53 + .62 + 3.14 = 10.29$$

The overall chi-square statistic is 10.29, but we need to determine whether this is a statistically significant number. To do this, we have to use the chi-square table in Appendix A of this book.

Using the Chi-Square Table

If the null hypothesis is true and there is no difference between the expected and observed frequencies, the chi-square value will be 0. If there is a difference in the expected and observed values, then the obtained chi-square will be greater than 0. The question is how much greater than 0 must the obtained chi-square value be for us to reject the null hypothesis? This is determined by comparing the calculated chi-square value to the critical chi-square value in the chi-square table.

One important difference between the chi-square table (a portion of which is reproduced in Table 15.1) and some of the other statistical tables we have used is that with the chi-square test we do not have to determine if we have a one- or a two-tailed test. All chi-square tests are essentially one tailed because our concern is whether there is a difference between categories. With a nominal-level variable, there is a difference or there is not a difference, which means that there is only one region of rejection. In addition, since we are calculating the chi-square from squared differences, the chi-square statistic will never be a negative number.

TABLE 15.1 The Chi-Square Table

df	$\alpha = .05$	$\alpha = .01$
1	3.84	6.64
2	5.99	9.21
3	7.81	11.34
4	9.49	13.28
5	11.07	15.09
6	12.59	16.81
7	14.07	18.48
8	15.51	20.09
9	16.92	21.67

The formula for the degrees of freedom for the one-way chi-square is

$$df = K - 1$$

where, again, K is the number of categories. In our problem, there are three categories (Yes, No, and Unsure), and so $3 - 1 = 2$. We have 2 degrees of freedom, and let us set alpha at .05, so the critical value for a chi-square (X^2_{crit}) with 2 degrees of freedom is 5.99. Our calculated chi-square (X^2_{obt}) of 10.29 lies beyond that, so it is in the region of rejection, which means we reject the null hypothesis and conclude that there are differences in voting preferences for Candidate A.

Since we have a significant difference in voting preferences, we conclude that these differences are not due to random chance but to actual differences among voters. Since this is an inferential procedure, we then conclude that since 38% of our sample indicated that they are going to vote for Candidate A (380/1,000), then 38% of the population will vote for Candidate A. We assume this because we used the proper sampling techniques, but remember that there is a .05 probability (or 5% chance) that we are wrong due to sampling error.

Now let us say we go to another city to do some political polling there. This time we randomly sample 1,500 people and ask them about their voting preference for Candidate B. We use the same categories as before and get the following results:

Will You Vote for Candidate B?

Yes		No		Unsure	
510	500	500	500	490	500

Our statistical hypotheses are the same as before, so

H_A: There are differences in voter preferences.

H_0: There are no differences in voter preferences.

The observed frequencies are 510 for Yes, 500 for No, and 490 for Unsure. The formula for the expected frequencies is N/K, so $1{,}500/3 = 500$. Remember, the expected frequencies are the frequencies we would expect if the null hypothesis is correct. So are the differences we see between the observed and expected frequencies enough that we conclude that they are significant, or are they due to random chance?

We have to calculate a chi-square for each cell, so for the Yes category

$$X^2 = \frac{(510 - 500)^2}{500} = \frac{(10)^2}{500} = \frac{100}{500} = .2$$

and for the No category

$$X^2 = \frac{(500 - 500)^2}{500} = \frac{(0)^2}{500} = \frac{0}{500} = 0$$

and finally for the Unsure category

$$X^2 = \frac{(490 - 500)^2}{500} = \frac{(-10)^2}{500} = \frac{100}{500} = .2$$

Our final step is to add the chi-square for each cell to give us the overall chi-square statistic

$$X^2 = .2 + 0 + .2 = .4$$

Our degrees of freedom $(K-1)$ have not changed, and so our critical value of chi-square (5.99) has not changed either. Since .4 is less than 5.99, it does not fall in the region of rejection, so we accept the null hypothesis and conclude that there are no differences in voter preference.

CALCULATING THE TWO-WAY CHI-SQUARE

With the one-way chi-square, we were examining one variable. The two-way chi-square examines two variables or, more specifically, whether one variable has an effect on another. If you recall in Chapter 1, we discussed independent and dependent variables. The independent variable is said to cause or have an effect on the dependent variable. We have already examined independent and dependent variables at the interval/ratio level. For example, in Chapter 12 we examined whether undergoing training (independent variable) causes people to assemble more desks (dependent variable) than someone who has not had the training. The two-way chi-square will do the same thing for variables measured at the nominal level.

© Leone_V/Shutterstock.com

Let us say we are interested in determining how people feel about our political candidates, but now we want to know whether gender has any effect on voter preference. In this case, gender is the independent variable and voter preference is the dependent variable. We will categorize the dependent variable the same way as before—Yes, No, and Unsure—but this time we will break our responses out by gender. We take a sample of 500 people, ask them how likely they are to vote for Candidate A, and get the following results:

	Will You Vote for Candidate A?			
	Yes	No	Unsure	Row Totals
Female	150	70	30	250
Male	100	80	70	250
Column Totals	250	150	100	

The statistical hypotheses with the two-way chi-square are different than the statistical hypotheses with one-way chi-square because of the presence of a second variable. The two-way chi-square is also called the test of independence because it is testing whether the frequencies in one category (answering yes to the question) depend on which category one belongs to in the other variable (being male or female). Therefore, the alternative hypothesis for a two-way chi-square is

H_A: Voter preference and gender are dependent.

This is stating that how likely someone is to vote for Candidate A **depends** on their gender. The null hypothesis is

H_0: Voter preference and gender are independent.

By independent, we are saying that the dependent variable (voter preference) is not related to the independent variable (gender). In the two-way chi-square, the statistical hypotheses always start with the variables under consideration and end with either "are dependent" (for the alternative hypothesis) or "are independent" (for the null hypothesis).

The numbers in the cells are the observed frequencies. We need to determine the expected frequencies (the frequencies we expect if the null hypothesis is correct), but we do this differently than what is done in the one-way chi-square. The formula for the expected frequencies in the two-way chi-square is

$$f_e = \frac{(\text{row total})(\text{column total})}{N}$$

Row total is the total observed frequencies for each row in the table. Column total is the total observed frequencies for each column in the table. N is our sample size.

In our example, there are 150 women who say they will vote for Candidate A, 70 women who say they will not, and 30 women who are unsure. So we add these together to get the total for the female row. We have to do this for each cell. So, to get the expected frequencies for females who said yes we multiply 250 (the row total for that cell) times 250 (the column total for that cell) and get 62,500. We then divide by 500 (the sample size) and get an expected frequency of 125. This means that if the null hypothesis is correct, we would expect 125 women to say that they would vote for Candidate A. We do this for each cell (in bold)

Women No $f_e = (250)(150)/500 = 75$

Women Unsure $f_e = (250)(100)/500 = 50$

Men Yes $f_e = (250)(250)/500 = 125$

Men No $f_e = (250)(150)/500 = 75$

Men Unsure $f_e = (250)(100)/500 = 50$

The table is reproduced below without the row and column totals but with the expected frequencies in each cell (in bold)

	Will You Vote for Candidate A?					
	Yes		No		Unsure	
Female	150	**125**	70	**75**	30	**50**
Male	100	**125**	80	**75**	70	**50**

The formula for the two-way chi-square is the same as the one-way chi-square

$$X^2 = \sum \left(\frac{(f_o - f_e)^2}{f_e} \right)$$

We have to calculate $\frac{(f_o - f_e)^2}{f_e}$ for each cell, then add the results together. So, for the cell of women who said Yes

$$X^2 = \frac{(150 - 125)^2}{125} = \frac{(25)^2}{125} = \frac{625}{125} = 5$$

the chi-square is 5. We then do this for each of the other cells

Women who said No: $X^2 = \dfrac{(80 - 75)^2}{75} = \dfrac{(5)^2}{75} = \dfrac{25}{75} = .33$

Women who are Unsure: $X^2 = \dfrac{(30 - 50)^2}{50} = \dfrac{(-20)^2}{50} = \dfrac{400}{50} = 8$

Men who said Yes: $X^2 = \dfrac{(100 - 125)^2}{125} = \dfrac{(-25)^2}{125} = \dfrac{625}{125} = 5$

Men who said No: $X^2 = \dfrac{(80 - 75)^2}{75} = \dfrac{(5)^2}{75} = \dfrac{25}{75} = .33$

Men who are Unsure: $X^2 = \dfrac{(70 - 50)^2}{50} = \dfrac{(20)^2}{50} = \dfrac{400}{50} = 8$

Our last step is to sum all of the results for each cell to get our overall chi-square statistic

$$X^2 = 5 + .33 + 8 + 5 + .33 + 8 = 26.66$$

Once again, to determine whether this is statistically significant we compare this number to the critical values of chi-square found in Appendix A. However, the formula for the degrees of freedom for the two-way chi-square is different than the formula for the degrees of freedom for the one-way chi-square

$$df = (\text{number of rows} - 1)(\text{number of columns} - 1)$$

In this problem, we have two rows (male and female) and three columns (Yes, No, and Unsure), so our degrees of freedom are

$$df = (2-1)(3-1)$$

$$df = (1)(2)$$

$$df = 2$$

With alpha set at .05, the critical value for chi-square (X^2_{crit}) with 2 degrees of freedom is 5.99. Our calculated chi-square (X^2_{obt}) of 26.66 lies beyond that, so it is in the region of rejection. This means we reject the null hypothesis and conclude that gender and voter preference are dependent; gender does have an effect on voter preference (at least in regard to Candidate A).

Let us look at Candidate B to see if gender has an effect on voter preference. We survey 1,000 people, break the responses out by gender, and get the following results

	Will You Vote for Candidate B?			
	Yes	No	Unsure	Row Totals
Female	175	200	125	500
Male	180	210	110	500
Column Totals	355	410	235	

Our statistical hypotheses are the same as the previous problem

H_A: Voter preference and gender are dependent.

H_0: Voter preference and gender are independent.

We use the row and column totals to determine the expected frequencies for each cell. Remember, the formula is

$$f_e = \frac{(\text{row total})(\text{column total})}{N}$$

so the expected frequencies are as follows:

Women Yes: $f_e = (500)(355)/1000 = 177.5$

Women No $f_e = (500)(410)/1000 = 205$

Women Unsure $f_e = (500)(235)/1000 = 117.5$

Men Yes $f_e = (500)(355)/1000 = 177.5$

Men No $f_e = (500)(410)/1000 = 205$

Men Unsure $f_e = (500)(235)/1000 = 117.5$

Here is the table with the expected frequencies

	Will You Vote for Candidate B?					
	Yes		No		Unsure	
Female	175	177.5	200	205	125	117.5
Male	180	177.5	210	205	110	117.5

Now we can calculate the chi-square statistic for each cell

$$\text{Women who said Yes: } X^2 = \frac{(175 - 177.5)^2}{177.5} = \frac{(-2.5)^2}{177.5} = \frac{6.25}{177.5} = .04$$

$$\text{Women who said No: } X^2 = \frac{(200 - 205)^2}{205} = \frac{(-5)^2}{205} = \frac{25}{205} = .12$$

$$\text{Women who are Unsure: } X^2 = \frac{(125 - 117.5)^2}{117.5} = \frac{(-7.5)^2}{117.5} = \frac{56.25}{117.5} = .48$$

$$\text{Men who said Yes: } X^2 = \frac{(180 - 177.5)^2}{177.5} = \frac{(2.5)^2}{177.5} = \frac{6.25}{177.5} = .04$$

$$\text{Men who said No: } X^2 = \frac{(210 - 205)^2}{205} = \frac{(5)^2}{205} = \frac{25}{205} = .12$$

$$\text{Men who are Unsure: } X^2 = \frac{(110 - 117.5)^2}{117.5} = \frac{(-7.5)^2}{117.5} = \frac{56.25}{117.5} = .48$$

Our last step is to sum all of the results for each cell to get our overall chi-square statistic

$$X^2 = .04 + .12 + .48 + .04 + .12 + .48 = 1.28$$

We have the same degrees of freedom (2) as before because we have the same number of rows and columns, so the critical chi-square value is still 5.99. Since our obtained chi-square value is lower than that, it does not fall in the region of rejection, so we retain the null hypothesis. Gender has no effect on voter preference for Candidate B.

MEASURING EFFECT SIZE

There is a post hoc test with the two-way chi-square that measures effect size. Like the previous post hoc tests, we only perform the post hoc test for the two-way chi-square when we reject the null hypothesis. The post hoc test we use is called the phi coefficient (Φ), and it measures effect size. In our first problem with the two-way chi-square, we saw that gender does have an effect on voter preference for Candidate A. The phi coefficient answers the question, "how much of an effect?" The phi coefficient will be a number between 0 (no effect) and 1 (very strong effect), and the formula is

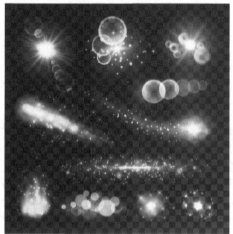

$$\Phi = \sqrt{\frac{X^2_{obt}}{N}}$$

Earlier, we saw that gender has an effect on voter preference for Candidate A. The obtained (calculated) chi-square was 26.66 and our sample size was 500, so

$$\Phi = \sqrt{\frac{26.66}{500}}$$

$$\Phi = \sqrt{.05}$$

$$\Phi = .22$$

We interpret Φ the same way we interpret the Pearson's r correlation coefficient (see Chapter 7). So while gender does have an effect on voter preference, it is a weak relationship. We can also square Φ to determine predictive ability (just like r^2). So if we square .22, we get .05. Since gender has a weak effect on voter preference, we only improve our predictive ability by 5% if we know the gender of the person voting for Candidate A than if we do not. One other way to interpret these findings is that 22% of the variance in voter preference is attributable to gender.

CHAPTER SUMMARY

The chi-square test is a nonparametric inferential statistic. Nonparametric inferential statistics are used when our data do not meet the assumptions for parametric procedures. The chi-square test is used when the dependent variable is measured at the nominal level. We use a one-way chi-square when we are examining one variable and the two-way chi-square when we want to see if one variable has an effect on another variable. If an effect is found, we can measure the size of the effect by calculating the phi coefficient.

FORMULAS FOR CHAPTER 15

Chi-Square: $X^2 = \sum \left(\dfrac{(f_o - f_e)^2}{f_e} \right)$

One-way Chi-Square: $f_e = N/K$

Two-way Chi-Square: $f_e = \dfrac{(\text{row total})(\text{column total})}{N}$

One-way chi-square: $df = K-1$

Two-way chi-square: $df = (\text{number of rows} - 1)(\text{number of columns} - 1)$

phi coefficient: $\Phi = \sqrt{\dfrac{X^2_{\text{obt}}}{N}}$

PRACTICE PROBLEMS

1. We are interested in attitude toward the death penalty and ask 600 people if they support the death penalty, do not support the death penalty, or have no opinion. The results are as follows:

Support	Do Not Support	No Opinion
310	190	100

   a. What is the null hypothesis?
   b. What is the alternative hypothesis?
   c. With $\alpha = .05$, what is the critical chi-square?
   d. What is your chi-square statistic?
   e. Is there a significant difference between attitudes toward the death penalty? If so, what percentage of the population would be expected to respond to the question the same way?

2. We are interested in the relationship between gender and attitude towards the death penalty.

Attitude Toward the Death Penalty	Gender	
	Male	Female
Approve	28	32
Disapprove	30	35
Neutral	20	19

   a. What is the independent variable?
   b. What is the dependent variable?
   c. What is the null hypothesis?
   d. What is the alternative hypothesis?
   e. With $\alpha = .05$, what is the critical chi-square?
   f. What is your chi-square statistic?
   g. Does gender have an effect on attitude toward the death penalty? If so, how strong is it?

3. We are interested in whether there is a relationship between the type of institution where prison guards work and their satisfaction with the job.

Type of Institution	Satisfied with Job?	
	No	Yes
Medium	15	30
Maximum	100	40

a. What is the independent variable?
b. What is the dependent variable?
c. What is the null hypothesis?
d. What is the alternative hypothesis?
e. With $\alpha = .05$, what is the critical chi-square?
f. What is your chi-square statistic?
g. Does type of institution have an effect on job satisfaction? If so, how strong is it?

4. We are interested in whether year in school (undergraduate or graduate) has effect on support for a college's drinking policy.

Campus Drinking Policy	Year in School	
	Undergraduate	Graduate
Support	150	75
Oppose	200	90

a. What is the independent variable?
b. What is the dependent variable?
c. What is the null hypothesis?
d. What is the alternative hypothesis?
e. With $\alpha = .05$, what is the critical chi-square?
f. What is your chi-square statistic?
g. Does year in school have an effect on support for the drinking policy? If so, how strong is it?

5. The manager of a restaurant is purchasing a jukebox and wants to know patrons' preferences for different types of music. She randomly samples 200 people, asks them their favorite type of music, and gets the following results

New Wave	Heavy Metal	Dance
80	50	70

   a. What is the null hypothesis?
   b. What is the alternative hypothesis?
   c. With $\alpha = .05$, what is the critical chi-square?
   d. What is your chi-square statistic?
   e. Is there a significant difference between types of music liked? If so, what percentage of the population would be expected to respond to the question the same way?

6. The restaurant manager then wants to see if gender has an effect on musical preferences so she randomly samples 200 people, asks them their favorite type of music, and gets the following results

Favorite Type of Music	Gender	
	Male	Female
New Wave	50	30
Heavy Metal	25	25
Dance	25	45

   a. What is the independent variable?
   b. What is the dependent variable?
   c. What is the null hypothesis?
   d. What is the alternative hypothesis?
   e. With $\alpha = .05$, what is the critical chi-square?
   f. What is your chi-square statistic?
   g. Does gender have an effect on musical preference? If so, how strong is it?

STATISTIC TABLES

TABLE 1 Proportional Areas under the Standard Normal Curve: The *Z* Tables

Column A contains the absolute value of the Z score.
Column B contains the proportion of the area of the normal curve between the mean ($\mu = 0$) and the Z score.
Column C contains the proportion of the area of the normal curve from the Z score to the tail of the curve.

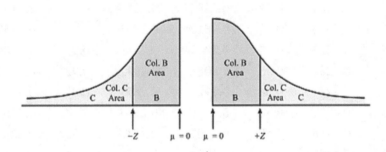

±Z A	B	C	±Z A	B	C	±Z A	B	C
0.00	0.0000	0.5000	0.37	0.1443	0.3557	0.74	0.2704	0.2296
0.01	0.0040	0.4960	0.38	0.1480	0.3520	0.75	0.2734	0.2266
0.02	0.0080	0.4920	0.39	0.1517	0.3483	0.76	0.2764	0.2236
0.03	0.0120	0.4880	0.40	0.1554	0.3446	0.77	0.2794	0.2206
0.04	0.0160	0.4840	0.41	0.1591	0.3409	0.78	0.2823	0.2177
0.05	0.0199	0.4801	0.42	0.1628	0.3372	0.79	0.2852	0.2148
0.06	0.0239	0.4761	0.43	0.1664	0.3336	0.80	0.2881	0.2119
0.07	0.0279	0.4721	0.44	0.1700	0.3300	0.81	0.2910	0.2090
0.08	0.0319	0.4681	0.45	0.1736	0.3264	0.82	0.2939	0.2061
0.09	0.0359	0.4641	0.46	0.1772	0.3228	0.83	0.2967	0.2033
0.10	0.0398	0.4602	0.47	0.1808	0.3192	0.84	0.2995	0.2005
0.11	0.0438	0.4562	0.48	0.1844	0.3156	0.85	0.3023	0.1977
0.12	0.0478	0.4522	0.49	0.1879	0.3121	0.86	0.3051	0.1949
0.13	0.0517	0.4483	0.50	0.1915	0.3085	0.87	0.3078	0.1922
0.14	0.0557	0.4443	0.51	0.1950	0.3050	0.88	0.3106	0.1894
0.15	0.0596	0.4404	0.52	0.1985	0.3015	0.89	0.3133	0.1867
0.16	0.0636	0.4364	0.53	0.2019	0.2981	0.90	0.3159	0.1841
0.17	0.0675	0.4325	0.54	0.2054	0.2946	0.91	0.3186	0.1814
0.18	0.0714	0.4286	0.55	0.2088	0.2912	0.92	0.3212	0.1788
0.19	0.0753	0.4247	0.56	0.2123	0.2877	0.93	0.3238	0.1762
0.20	0.0793	0.4207	0.57	0.2157	0.2843	0.94	0.3264	0.1736
0.21	0.0832	0.4168	0.58	0.2190	0.2810	0.95	0.3289	0.1711
0.22	0.0871	0.4129	0.59	0.2224	0.2776	0.96	0.3315	0.1685
0.23	0.0910	0.4090	0.60	0.2257	0.2743	0.97	0.3340	0.1660
0.24	0.0948	0.4052	0.61	0.2291	0.2709	0.98	0.3365	0.1635
0.25	0.0987	0.4013	0.62	0.2324	0.2676	0.99	0.3389	0.1611
0.26	0.1026	0.3974	0.63	0.2357	0.2643	1.00	0.3413	0.1587
0.27	0.1064	0.3936	0.64	0.2389	0.2611	1.01	0.343	0.1562
0.28	0.1103	0.3897	0.65	0.2422	0.2578	1.02	0.3461	0.1539
0.29	0.1141	0.3859	0.66	0.2454	0.2546	1.03	0.3485	0.1515
0.30	0.1179	0.3821	0.67	0.2486	0.2514	1.04	0.3508	0.1492
0.31	0.1217	0.3783	0.68	0.2517	0.2483	1.05	0.3531	0.1469
0.32	0.1255	0.3745	0.69	0.2549	0.2451	1.06	0.3554	0.1446
0.33	0.1293	0.3707	0.70	0.2580	0.2420	1.07	0.3577	0.1423
0.34	0.1331	0.3669	0.71	0.2611	0.2389	1.08	0.3599	0.1401
0.35	0.1368	0.3632	0.72	0.2642	0.2358	1.09	0.3621	0.1379
0.36	0.1406	0.3594	0.73	0.2673	0.2327	1.10	0.3643	0.1357

±Z A	B	C	±Z A	B	C	±Z A	B	C
1.11	0.3665	0.1335	1.48	0.4306	0.0694	1.84	0.4671	0.0329
1.12	0.3686	0.1314	1.49	0.4319	0.0681	1.85	0.4678	0.0322
1.13	0.3708	0.1292	1.50	0.4332	0.0668	1.86	0.4686	0.0314
1.14	0.3729	0.1271	1.51	0.4345	0.0665	1.87	0.4693	0.0307
1.15	0.3749	0.1251	1.52	0.4357	0.0643	1.88	0.4699	0.0301
1.16	0.3770	0.1230	1.53	0.4370	0.0630	1.89	0.4706	0.0294
1.17	0.3790	0.1210	1.54	0.4382	0.0618	1.90	0.4713	0.0287
1.18	0.3810	0.1190	1.55	0.4394	0.0606	1.91	0.4719	0.0281
1.19	0.3830	0.1170	1.56	0.4406	0.0594	1.92	0.4726	0.0274
1.20	0.3849	0.1151	1.57	0.4418	0.0582	1.93	0.4732	0.0268
1.21	0.3869	0.1131	1.58	0.4429	0.0571	1.94	0.4738	0.0262
1.22	0.3888	0.1112	1.59	0.4441	0.0559	1.95	0.4744	0.0256
1.23	0.3907	0.1093	1.60	0.4452	0.0548	1.96	0.4750	0.0250
1.24	0.3925	0.1075	1.61	0.4463	0.0537	1.97	0.4756	0.0244
1.25	0.3944	0.1056	1.62	0.4474	0.0526	1.98	0.4761	0.0239
1.26	0.3962	0.1038	1.63	0.4484	0.0516	1.99	0.4767	0.0233
1.27	0.3980	0.1020	1.64	0.4495	0.0505	2.00	0.4772	0.0228
1.28	0.3997	0.1003	1.645	0.4500	0.0500	2.01	0.4778	0.0222
1.29	0.4015	0.0985	1.65	0.4505	0.0495	2.02	0.4783	0.0217
1.30	0.4032	0.0968	1.66	0.4515	0.0485	2.03	0.4788	0.0212
1.31	0.4049	0.0951	1.67	0.4525	0.0475	2.04	0.4793	0.0207
1.32	0.4066	0.0934	1.68	0.4535	0.0465	2.05	0.4798	0.0202
1.33	0.4082	0.0918	1.69	0.4545	0.0455	2.06	0.4803	0.0197
1.34	0.4099	0.0901	1.70	0.4554	0.0446	2.07	0.4808	0.0192
1.35	0.4115	0.0885	1.71	0.4564	0.0436	2.08	0.4812	0.0188
1.36	0.4131	0.0869	1.72	0.4573	0.0427	2.09	0.4817	0.0183
1.37	0.4147	0.0853	1.73	0.4582	0.0418	2.10	0.4821	0.0179
1.38	0.4162	0.0838	1.74	0.4591	0.0409	2.11	0.4826	0.0174
1.39	0.4177	0.0823	1.75	0.4599	0.0401	2.12	0.4830	0.0170
1.40	0.4192	0.0808	1.76	0.4608	0.0392	2.13	0.4834	0.0166
1.41	0.4207	0.0793	1.77	0.4616	0.0384	2.14	0.4838	0.0162
1.42	0.4222	0.0778	1.78	0.4625	0.0375	2.15	0.4842	0.0158
1.43	0.4236	0.0764	1.79	0.4633	0.0367	2.16	0.4846	0.0154
1.44	0.4251	0.0749	1.80	0.4641	0.0359	2.17	0.4850	0.0150
1.45	0.4265	0.0735	1.81	0.4649	0.0351	2.18	0.4854	0.0146
1.46	0.4279	0.0721	1.82	0.4656	0.0344	2.19	0.4857	0.0143
1.47	0.4292	0.0708	1.83	0.4664	0.0336	2.20	0.4861	0.0139

±Z A	B	C	±Z A	B	C	±Z A	B	C
2.21	0.4864	0.0136	2.56	0.4958	0.0052	2.90	0.4981	0.0019
2.22	0.4868	0.0132	2.57	0.4949	0.0051	2.91	0.4982	0.0018
2.23	0.4871	0.0129	2.575	0.4950	0.0050	2.92	0.4983	0.0017
2.24	0.4875	0.0125	2.58	0.4951	0.0049	2.93	0.4983	0.0017
2.25	0.4878	0.0122	2.59	0.4952	0.0048	2.94	0.4984	0.0016
2.26	0.4881	0.0119	2.60	0.4953	0.0047	2.95	0.4984	0.0016
2.27	0.4884	0.0116	2.61	0.4955	0.0045	2.96	0.4985	0.0015
2.28	0.4887	0.0113	2.62	0.4956	0.0044	2.97	0.4985	0.0015
2.29	0.4890	0.0110	2.63	0.4957	0.0043	2.98	0.4986	0.0014
2.30	0.4893	0.0107	2.64	0.4959	0.0041	2.99	0.4986	0.0014
2.31	0.4896	0.0104	2.65	0.4960	0.0040	3.00	0.4987	0.0013
2.32	0.4898	0.0102	2.66	0.4961	0.0039	3.01	0.4987	0.0013
2.33	0.4901	0.0099	2.67	0.4962	0.0038	3.02	0.4987	0.0013
2.34	0.4904	0.0096	2.68	0.4963	0.0037	3.03	0.4988	0.0012
2.35	0.4906	0.0094	2.69	0.4964	0.0036	3.04	0.4988	0.0012
2.36	0.4909	0.0091	2.70	0.4965	0.0035	3.05	0.4989	0.0011
2.37	0.4911	0.0089	2.71	0.4966	0.0034	3.06	0.4989	0.0011
2.38	0.4913	0.0087	2.72	0.4967	0.0033	3.07	0.4989	0.0011
2.39	0.4916	0.0084	2.73	0.4968	0.0032	3.08	0.4990	0.0010
2.40	0.4918	0.0082	2.74	0.4969	0.0031	3.09	0.4990	0.0010
2.41	0.4920	0.0080	2.75	0.4970	0.0030	3.10	0.4990	0.0010
2.42	0.4922	0.0078	2.76	0.4971	0.0029	3.11	0.4991	0.0009
2.43	0.4925	0.0075	2.77	0.4972	0.0028	3.12	0.4991	0.0009
2.44	0.4927	0.0073	2.78	0.4973	0.0027	3.13	0.4991	0.0009
2.45	0.4929	0.0071	2.79	0.4974	0.0026	3.14	0.4992	0.0008
2.46	0.4931	0.0069	2.80	0.4974	0.0026	3.15	0.4992	0.0008
2.47	0.4932	0.0068	2.81	0.4975	0.0025	3.16	0.4992	0.0008
2.48	0.4934	0.0066	2.82	0.4976	0.0024	3.17	0.4992	0.0008
2.49	0.4936	0.0064	2.83	0.4977	0.0023	3.18	0.4993	0.0007
2.50	0.4938	0.0062	2.84	0.4977	0.0023	3.19	0.4993	0.0007
2.51	0.4940	0.0060	2.85	0.4978	0.0022	3.20	0.4993	0.0007
2.52	0.4941	0.0059	2.86	0.4979	0.0021	3.21	0.4993	0.0007
2.53	0.4943	0.0057	2.87	0.4980	0.0020	3.22	0.4994	0.0006
2.54	0.4945	0.0055	2.88	0.4980	0.0020	3.23	0.4994	0.0006
2.55	0.4946	0.0054	2.89	0.4981	0.0019	3.24	0.4994	0.0006

| ±Z | | | ±Z | | | ±Z | | |
A	B	C	A	B	C	A	B	C
3.25	0.4994	0.0006	3.29	0.4994	0.0006	3.60	0.4998	0.0002
3.26	0.4994	0.0006	3.30	0.4995	0.0005	3.70	0.4999	0.0001
3.27	0.4994	0.0006	3.40	0.4997	0.0003			
3.28	0.4994	0.0006	3.50	0.4998	0.0008			

TABLE 2 Critical Values of the Chi Square: The χ^2 Tables

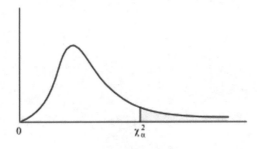

Level of Significance

df	$\alpha = 0.05$	$\alpha = 0.01$
1	3.84	6.64
2	5.99	9.21
3	7.81	11.34
4	9.49	13.28
5	11.07	15.09
6	12.59	16.81
7	14.07	18.48
8	15.51	20.09
9	16.92	21.67
10	18.31	23.21
11	19.68	24.73
12	21.03	26.22
13	22.36	27.69
14	23.68	29.14
15	25.00	30.58
16	26.30	32.00
17	27.59	33.41
18	28.87	34.81
19	30.14	36.19
20	31.41	37.57
25	37.65	44.31
30	43.77	50.89
35	49.80	57.34
40	55.76	63.69
50	67.50	76.15
75	96.22	106.39
100	124.34	135.81

Source: Table prepared by the author using Microsoft Excel's chi inverse function. (Excel is a registered trademark of the Microsoft Corporation.)

From *Statistical Analysis in The Behavioral Sciences* by James Raymondo. Copyright © 2015 by Kendall Hunt Publishing Company. Reprinted by permission

TABLE 3 Critical Values of *t* for Two-Tail and One-Tail Tests: The *t* Tables

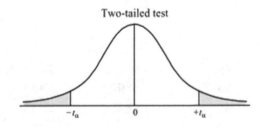

Two-tailed test

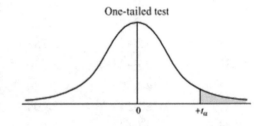

One-tailed test

	Level of Significance				Level of Significance	
df	$\alpha = 0.05$	$\alpha = 0.01$		*df*	$\alpha = 0.05$	$\alpha = 0.01$
1	12.706	63.656		1	6.314	31.821
2	4.303	9.925		2	2.920	6.965
3	3.182	5.841		3	2.353	4.541
4	2.776	4.604		4	2.132	3.747
5	2.571	4.032		5	2.015	3.365
6	2.447	3.707		6	1.943	3.143
7	2.365	3.499		7	1.895	2.998
8	2.306	3.355		8	1.860	2.896
9	2.262	3.250		9	1.833	2.821
10	2.228	3.169		10	1.812	2.764
11	2.201	3.106		11	1.796	2.718
12	2.179	3.055		12	1.782	2.681
13	2.160	3.012		13	1.771	2.650
14	2.145	2.977		14	1.761	2.624
15	2.131	2.947		15	1.753	2.602
16	2.120	2.921		16	1.746	2.583
17	2.110	2.898		17	1.740	2.567
18	2.101	2.878		18	1.734	2.552
19	2.093	2.861		19	1.729	2.539
20	2.086	2.845		20	1.725	2.528
21	2.080	2.831		21	1.721	2.518
22	2.074	2.819		22	1.717	2.508
23	2.069	2.807		23	1.714	2.500
24	2.064	2.797		24	1.711	2.492
25	2.060	2.787		25	1.708	2.485
26	2.056	2.779		26	1.706	2.479

df	$\alpha = 0.05$	$\alpha = 0.01$	df	$\alpha = 0.05$	$\alpha = 0.01$
27	2.052	2.771	27	1.703	2.473
28	2.048	2.763	28	1.701	2.467
29	2.045	2.756	29	1.699	2.462
30	2.042	2.750	30	1.697	2.457
35	2.030	2.724	35	1.690	2.438
40	2.021	2.704	40	1.684	2.423
45	2.014	2.690	45	1.679	2.412
50	2.009	2.678	50	1.676	2.403
75	1.992	2.643	75	1.665	2.377
100	1.984	2.626	100	1.660	2.364
1000	1.962	2.581	1000	1.646	2.330
∞	1.960	2.576	∞	1.645	2.326

Source: Table prepared by the author using Microsoft Excel's *t* Inverse function. (Excel is a registered trademark of the Microsoft Corporation

From *Statistical Analysis in The Behavioral Sciences* by James Raymondo. Copyright © 2015 by Kendall Hunt Publishing Company. Reprinted by permission

Table 4 Critical Values of the *F* Statistic: The *F* Tables

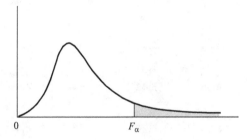

Critical values for α = .05 in regular text
Critical values for α = .01 in **bold text**

Degrees of Freedom within Groups (df_{wn} = N − k)	α	Degrees of Freedom Between Groups (df_{bn} = k − 1)						
		1	2	3	4	5	6	7
1	.05	161	200	216	225	230	234	237
	.01	**4052**	**4999**	**5404**	**5624**	**5764**	**5859**	**5928**
2	.05	18.51	19.00	19.16	19.25	19.30	19.33	19.35
	.01	**98.50**	**99.00**	**99.16**	**99.25**	**99.30**	**99.33**	**99.36**
3	.05	10.13	9.55	9.28	9.12	9.01	8.94	8.89
	.01	**34.12**	**30.82**	**29.46**	**28.71**	**28.24**	**27.91**	**27.67**
4	.05	7.71	6.94	6.59	6.39	6.26	6.16	6.09
	.01	**21.20**	**18.00**	**16.69**	**15.98**	**15.52**	**15.21**	**14.98**
5	.05	6.61	5.79	5.41	5.19	5.05	4.95	4.88
	.01	**16.26**	**13.27**	**12.06**	**11.39**	**10.97**	**10.67**	**10.46**
6	.05	5.99	5.14	4.76	4.53	4.39	4.28	4.21
	.01	**13.75**	**10.92**	**9.78**	**9.15**	**8.75**	**8.47**	**8.26**
7	.05	5.59	4.74	4.35	4.12	3.97	3.87	3.79
	.01	**12.25**	**9.55**	**8.45**	**7.85**	**7.46**	**7.19**	**6.99**
8	.05	5.32	4.46	4.07	3.84	3.69	3.58	3.50
	.01	**11.26**	**8.65**	**7.59**	**7.01**	**6.63**	**6.37**	**6.18**
9	.05	5.12	4.26	3.86	3.63	3.48	3.37	3.29
	.01	**10.56**	**8.02**	**6.99**	**6.42**	**6.06**	**5.80**	**5.61**
10	.05	4.96	4.10	3.71	3.48	3.33	3.22	3.14
	.01	**10.04**	**7.56**	**6.55**	**5.99**	**5.64**	**5.39**	**5.20**
11	.05	4.84	3.98	3.59	3.36	3.20	3.09	3.01
	.01	**9.65**	**7.21**	**6.22**	**5.67**	**5.32**	**5.07**	**4.89**
12	.05	4.75	3.89	3.49	3.26	3.11	3.00	2.91
	.01	**9.33**	**6.93**	**5.95**	**5.41**	**5.06**	**4.82**	**4.64**

	α	1	2	3	4	5	6	7
13	.05	4.67	3.81	3.41	3.18	3.03	2.92	2.83
	.01	9.07	6.70	5.74	5.21	4.86	4.62	4.44
14	.05	4.60	3.74	3.34	3.11	2.96	2.85	2.76
	.01	8.86	6.51	5.56	5.04	4.69	4.46	4.28
15	.05	4.54	3.68	3.29	3.06	2.90	2.79	2.71
	.01	8.68	6.36	5.42	4.89	4.56	4.32	4.14
16	.05	4.49	3.63	3.24	3.01	2.85	2.74	2.66
	.01	8.53	6.23	5.29	4.77	4.44	4.20	4.03
17	.05	4.45	3.59	3.20	2.96	2.81	2.70	2.61
	.01	8.40	6.11	5.19	4.67	4.34	4.10	3.93
18	.05	4.41	3.55	3.16	2.93	2.77	2.66	2.58
	.01	8.29	6.01	5.09	4.58	4.25	4.01	3.84
19	.05	4.38	3.52	3.13	2.90	2.74	2.63	2.54
	.01	8.18	5.93	5.01	4.50	4.17	3.94	3.77
20	.05	4.35	3.49	3.10	2.87	2.71	2.60	2.51
	.01	8.10	5.85	4.94	4.43	4.10	3.87	3.70
21	.05	4.32	3.47	3.07	2.84	2.68	2.57	2.49
	.01	8.02	5.78	4.87	4.37	4.04	3.81	3.64
22	.05	4.30	3.44	3.05	2.82	2.66	2.55	2.46
	.01	7.95	5.72	4.82	4.31	3.99	3.76	3.59
23	.05	4.28	3.42	3.03	2.80	2.64	2.53	2.44
	.01	7.88	5.66	4.76	4.26	3.94	3.71	3.54
24	.05	4.26	3.40	3.01	2.78	2.62	2.51	2.42
	.01	7.82	5.61	4.72	4.22	3.90	3.67	3.50
25	.05	4.24	3.39	2.99	2.76	2.60	2.49	2.40
	.01	7.77	5.57	4.68	4.18	3.85	3.63	3.46
26	.05	4.23	3.37	2.98	2.74	2.59	2.47	2.39
	.01	7.72	5.53	4.64	4.14	3.82	3.59	3.42
27	.05	4.21	3.35	2.96	2.73	2.57	2.46	2.37
	.01	7.68	5.49	4.60	4.11	3.78	3.56	3.39
28	.05	4.20	3.34	2.95	2.71	2.56	2.45	2.36
	.01	7.64	5.45	4.57	4.07	3.75	3.53	3.36
29	.05	4.18	3.33	2.93	2.70	2.55	2.43	2.35
	.01	7.60	5.42	4.54	4.04	3.73	3.50	3.33
30	.05	4.17	3.32	2.92	2.69	2.53	2.42	2.33
	.01	7.56	5.39	4.51	4.02	3.70	3.47	3.30

	α	1	2	3	4	5	6	7
35	.05	4.12	3.27	2.87	2.64	2.49	2.37	2.29
	.01	**7.42**	**5.27**	**4.40**	**3.91**	**3.59**	**3.37**	**3.20**
40	.05	4.08	3.23	2.84	2.61	2.45	2.34	2.25
	.01	**7.31**	**5.18**	**4.31**	**3.83**	**3.51**	**3.29**	**3.12**
45	.05	4.06	3.20	2.81	2.58	2.42	2.31	2.22
	.01	**7.23**	**5.11**	**4.25**	**3.77**	**3.45**	**3.23**	**3.07**
50	.05	4.03	3.18	2.79	2.56	2.40	2.29	2.20
	.01	**7.17**	**5.06**	**4.20**	**3.72**	**3.41**	**3.19**	**3.02**
75	.05	3.97	3.12	2.73	2.49	2.34	2.22	2.13
	.01	**6.99**	**4.90**	**4.05**	**3.58**	**3.27**	**3.05**	**2.89**
100	.05	3.94	3.09	2.70	2.46	2.31	2.19	2.10
	.01	**6.90**	**4.82**	**3.98**	**3.51**	**3.21**	**2.99**	**2.82**
250	.05	3.88	3.03	2.64	2.41	2.25	2.13	2.05
	.01	**6.74**	**4.69**	**3.86**	**3.40**	**3.09**	**2.87**	**2.71**
500	.05	3.86	3.01	2.62	2.39	2.23	2.12	2.03
	.01	**6.69**	**4.65**	**3.82**	**3.36**	**3.05**	**2.84**	**2.68**
1000	.05	3.85	3.00	2.61	2.38	2.22	2.11	2.02
	.01	**6.66**	**4.63**	**3.80**	**3.34**	**3.04**	**2.82**	**2.66**

Critical values for α = .05 in regular text
Critical values for α = .01 in **bold text**

Degrees of Freedom within Groups (df_{wn} = N − k)	α	Degrees of Freedom Between Groups (df_{bn} = k − 1)						
		8	9	10	11	12	13	14
1	.05	239	241	242	243	244	245	245
	.01	**5981**	**6022**	**6056**	**6083**	**6107**	**6126**	**6143**
2	.05	19.37	19.38	19.40	19.40	19.41	19.42	19.42
	.01	**99.38**	**99.39**	**99.40**	**99.41**	**99.42**	**99.42**	**99.43**
3	.05	8.85	8.81	8.79	8.76	8.74	8.73	8.71
	.01	**27.49**	**27.34**	**27.23**	**27.13**	**27.05**	**26.98**	**26.92**
4	.05	6.04	6.00	5.96	5.94	5.91	5.89	5.87
	.01	**14.80**	**14.66**	**14.55**	**14.45**	**14.37**	**14.31**	**14.25**
5	.05	4.82	4.77	4.74	4.70	4.68	4.66	4.64
	.01	**10.29**	**10.16**	**10.05**	**9.96**	**9.89**	**9.82**	**9.77**
6	.05	4.15	4.10	4.06	4.03	4.00	3.98	3.96
	.01	**8.10**	**7.98**	**7.87**	**7.79**	**7.72**	**7.66**	**7.60**
7	.05	3.73	3.68	3.64	3.60	3.57	3.55	3.53
	.01	**6.84**	**6.72**	**6.62**	**6.54**	**6.47**	**6.41**	**6.36**
8	.05	3.44	3.39	3.35	3.31	3.28	3.26	3.24
	.01	**6.03**	**5.91**	**5.81**	**5.73**	**5.67**	**5.61**	**5.56**
9	.05	3.23	3.18	3.14	3.10	3.07	3.05	3.03
	.01	**5.47**	**5.35**	**5.26**	**5.18**	**5.11**	**5.05**	**5.01**
10	.05	3.07	3.02	2.98	2.94	2.91	2.89	2.86
	.01	**5.06**	**4.94**	**4.85**	**4.77**	**4.71**	**4.65**	**4.60**
11	.05	2.95	2.90	2.85	2.82	2.79	2.76	2.74
	.01	**4.74**	**4.63**	**4.54**	**4.46**	**4.40**	**4.34**	**4.29**
12	.05	2.85	2.80	2.75	2.72	2.69	2.66	2.64
	.01	**4.50**	**4.39**	**4.30**	**4.22**	**4.16**	**4.10**	**4.05**
13	.05	2.77	2.71	2.67	2.63	2.60	2.58	2.55
	.01	**4.30**	**4.19**	**4.10**	**4.02**	**3.96**	**3.91**	**3.86**
14	.05	2.70	2.65	2.60	2.57	2.53	2.51	2.48
	.01	**4.14**	**4.03**	**3.94**	**3.86**	**3.80**	**3.75**	**3.70**
15	.05	2.64	2.59	2.54	2.51	2.48	2.45	2.42
	.01	**4.00**	**3.89**	**3.80**	**3.73**	**3.67**	**3.61**	**3.56**
16	.05	2.59	2.54	2.49	2.46	2.42	2.40	2.37
	.01	**3.89**	**3.78**	**3.69**	**3.62**	**3.55**	**3.50**	**3.45**

	α	*8*	*9*	*10*	*11*	*12*	*13*	*14*
17	.05	2.55	2.49	2.45	2.41	2.38	2.35	2.33
	.01	**3.79**	**3.68**	**3.59**	**3.52**	**3.46**	**3.40**	**3.35**
18	.05	2.51	2.46	2.41	2.37	2.34	2.31	2.29
	.01	**3.71**	**3.60**	**3.51**	**3.43**	**3.37**	**3.32**	**3.27**
19	.05	2.48	2.42	2.38	2.34	2.31	2.28	2.26
	.01	**3.63**	**3.52**	**3.43**	**3.36**	**3.30**	**3.24**	**3.19**
20	.05	2.45	2.39	2.35	2.31	2.28	2.25	2.22
	.01	**3.56**	**3.46**	**3.37**	**3.29**	**3.23**	**3.18**	**3.13**
21	.05	2.42	2.37	2.32	2.28	2.25	2.22	2.20
	.01	**3.51**	**3.40**	**3.31**	**3.24**	**3.17**	**3.12**	**3.07**
22	.05	2.40	2.34	2.30	2.26	2.23	2.20	2.17
	.01	**3.45**	**3.35**	**3.26**	**3.18**	**3.12**	**3.07**	**3.02**
23	.05	2.37	2.32	2.27	2.24	2.20	2.18	2.15
	.01	**3.41**	**3.30**	**3.21**	**3.14**	**3.07**	**3.02**	**2.97**
24	.05	2.36	2.30	2.25	2.22	2.18	2.15	2.13
	.01	**3.36**	**3.26**	**3.17**	**3.09**	**3.03**	**2.98**	**2.93**
25	.05	2.34	2.28	2.24	2.20	2.16	2.14	2.11
	.01	**3.32**	**3.22**	**3.13**	**3.06**	**2.99**	**2.94**	**2.89**
26	.05	2.32	2.27	2.22	2.18	2.15	2.12	2.09
	.01	**3.29**	**3.18**	**3.09**	**3.02**	**2.96**	**2.90**	**2.86**
27	.05	2.31	2.25	2.20	2.17	2.13	2.10	2.08
	.01	**3.26**	**3.15**	**3.06**	**2.99**	**2.93**	**2.87**	**2.82**
28	.05	2.29	2.24	2.19	2.15	2.12	2.09	2.06
	.01	**3.23**	**3.12**	**3.03**	**2.96**	**2.90**	**2.84**	**2.79**
29	.05	2.28	2.22	2.18	2.14	2.10	2.08	2.05
	.01	**3.20**	**3.09**	**3.00**	**2.93**	**2.87**	**2.81**	**2.77**
30	.05	2.27	2.21	2.16	2.13	2.09	2.06	2.04
	.01	**3.17**	**3.07**	**2.98**	**2.91**	**2.84**	**2.79**	**2.74**
35	.05	2.22	2.16	2.11	2.07	2.04	2.01	1.99
	.01	**3.07**	**2.96**	**2.88**	**2.80**	**2.74**	**2.69**	**2.64**
40	.05	2.18	2.12	2.08	2.04	2.00	1.97	1.95
	.01	**2.99**	**2.89**	**2.80**	**2.73**	**2.66**	**2.61**	**2.56**
45	.05	2.15	2.10	2.05	2.01	1.97	1.94	1.92
	.01	**2.94**	**2.83**	**2.74**	**2.67**	**2.61**	**2.55**	**2.51**
50	.05	2.13	2.07	2.03	1.99	1.95	1.92	1.89
	.01	**2.89**	**2.78**	**2.70**	**2.63**	**2.56**	**2.51**	**2.46**

	α	8	9	10	11	12	13	14
75	.05	2.06	2.01	1.96	1.92	1.88	1.85	1.83
	.01	**2.76**	**2.65**	**2.57**	**2.49**	**2.43**	**2.38**	**2.33**
100	.05	2.03	1.97	1.93	1.89	1.85	1.82	1.79
	.01	**2.69**	**2.59**	**2.50**	**2.43**	**2.37**	**2.31**	**2.27**
250	.05	1.98	1.92	1.87	1.83	1.79	1.76	1.73
	.01	**2.58**	**2.48**	**2.39**	**2.32**	**2.26**	**2.20**	**2.15**
500	.05	1.96	1.90	1.85	1.81	1.77	1.74	1.71
	.01	**2.55**	**2.44**	**2.36**	**2.28**	**2.22**	**2.17**	**2.12**
1000	.05	1.95	1.89	1.84	1.80	1.76	1.73	1.70
	.01	**2.53**	**2.43**	**2.34**	**2.27**	**2.20**	**2.15**	**2.10**

Critical values for $\alpha = .05$ in regular text
Critical values for $\alpha = .01$ in **bold text**

Degrees of Freedom within groups ($df_{wn} = N - k$)	α	Degrees of Freedom Between Groups ($df_{bn} = k - 1$)					
		15	16	17	18	19	20
1	.05	246	246	247	247	248	248
	.01	**6157**	**6170**	**6181**	**6191**	**6201**	**6209**
2	.05	19.43	19.43	19.44	19.44	19.44	19.45
	.01	**99.43**	**99.44**	**99.44**	**99.44**	**99.45**	**99.45**
3	.05	8.70	8.69	8.68	8.67	8.67	8.66
	.01	**26.87**	**26.83**	**26.79**	**26.75**	**26.72**	**26.69**
4	.05	5.86	5.84	5.83	5.82	5.81	5.80
	.01	**14.20**	**14.15**	**14.11**	**14.08**	**14.05**	**14.02**
5	.05	4.62	4.60	4.59	4.58	4.57	4.56
	.01	**9.72**	**9.68**	**9.64**	**9.61**	**9.58**	**9.55**
6	.05	3.94	3.92	3.91	3.90	3.88	3.87
	.01	**7.56**	**7.52**	**7.48**	**7.45**	**7.42**	**7.40**
7	.05	3.51	3.49	3.48	3.47	3.46	3.44
	.01	**6.31**	**6.28**	**6.24**	**6.21**	**6.18**	**6.16**
8	.05	3.22	3.20	3.19	3.17	3.16	3.15
	.01	**5.52**	**5.48**	**5.44**	**5.41**	**5.38**	**5.36**
9	.05	3.01	2.99	2.97	2.96	2.95	2.94
	.01	**4.96**	**4.92**	**4.89**	**4.86**	**4.83**	**4.81**
10	.05	2.85	2.83	2.81	2.80	2.79	2.77
	.01	**4.56**	**4.52**	**4.49**	**4.46**	**4.43**	**4.41**
11	.05	2.72	2.70	2.69	2.67	2.66	2.65
	.01	**4.25**	**4.21**	**4.18**	**4.15**	**4.12**	**4.10**
12	.05	2.62	2.60	2.58	2.57	2.56	2.54
	.01	**4.01**	**3.97**	**3.94**	**3.91**	**3.88**	**3.86**
13	.05	2.53	2.51	2.50	2.48	2.47	2.46
	.01	**3.82**	**3.78**	**3.75**	**3.72**	**3.69**	**3.66**
14	.05	2.46	2.44	2.43	2.41	2.40	2.39
	.01	**3.66**	**3.62**	**3.59**	**3.56**	**3.53**	**3.51**
15	.05	2.40	2.38	2.37	2.35	2.34	2.33
	.01	**3.52**	**3.49**	**3.45**	**3.42**	**3.40**	**3.37**

	α	15	16	17	18	19	20
16	.05	2.35	2.33	2.32	2.30	2.29	2.28
	.01	**3.41**	**3.37**	**3.34**	**3.31**	**3.28**	**3.26**
17	.05	2.31	2.29	2.27	2.26	2.24	2.23
	.01	**3.31**	**3.27**	**3.24**	**3.21**	**3.19**	**3.16**
18	.05	2.27	2.25	2.23	2.22	2.20	2.19
	.01	**3.23**	**3.19**	**3.16**	**3.13**	**3.10**	**3.08**
19	.05	2.23	2.21	2.20	2.18	2.17	2.16
	.01	**3.15**	**3.12**	**3.08**	**3.05**	**3.03**	**3.00**
20	.05	2.20	2.18	2.17	2.15	2.14	2.12
	.01	**3.09**	**3.05**	**3.02**	**2.99**	**2.96**	**2.94**
21	.05	2.18	2.16	2.14	2.12	2.11	2.10
	.01	**3.03**	**2.99**	**2.96**	**2.93**	**2.90**	**2.88**
22	.05	2.15	2.13	2.11	2.10	2.08	2.07
	.01	**2.98**	**2.94**	**2.91**	**2.88**	**2.85**	**2.83**
23	.05	2.13	2.11	2.09	2.08	2.06	2.05
	.01	**2.93**	**2.89**	**2.86**	**2.83**	**2.80**	**2.78**
24	.05	2.11	2.09	2.07	2.05	2.04	2.03
	.01	**2.89**	**2.85**	**2.82**	**2.79**	**2.76**	**2.74**
25	.05	2.09	2.07	2.05	2.04	2.02	2.01
	.01	**2.85**	**2.81**	**2.78**	**2.75**	**2.72**	**2.70**
26	.05	2.07	2.05	2.03	2.02	2.00	1.99
	.01	**2.81**	**2.78**	**2.75**	**2.72**	**2.69**	**2.66**
27	.05	2.06	2.04	2.02	2.00	1.99	1.97
	.01	**2.78**	**2.75**	**2.71**	**2.68**	**2.66**	**2.63**
28	.05	2.04	2.02	2.00	1.99	1.97	1.96
	.01	**2.75**	**2.72**	**2.68**	**2.65**	**2.63**	**2.60**
29	.05	2.03	2.01	1.99	1.97	1.96	1.94
	.01	**2.73**	**2.69**	**2.66**	**2.63**	**2.60**	**2.57**
30	.05	2.01	1.99	1.98	1.96	1.95	1.93
	.01	**2.70**	**2.66**	**2.63**	**2.60**	**2.57**	**2.55**
35	.05	1.96	1.94	1.92	1.91	1.89	1.88
	.01	**2.60**	**2.56**	**2.53**	**2.50**	**2.47**	**2.44**
40	.05	1.92	1.90	1.89	1.87	1.85	1.84
	.01	**2.52**	**2.48**	**2.45**	**2.42**	**2.39**	**2.37**
45	.05	1.89	1.87	1.86	1.84	1.82	1.81
	.01	**2.46**	**2.43**	**2.39**	**2.36**	**2.34**	**2.31**

	α	15	16	17	18	19	20
50	.05	1.87	1.85	1.83	1.81	1.80	1.78
	.01	2.42	2.38	2.35	2.32	2.29	2.27
75	.05	1.80	1.78	1.76	1.74	1.73	1.71
	.01	2.29	2.25	2.22	2.18	2.16	2.13
100	.05	1.77	1.75	1.73	1.71	1.69	1.68
	.01	2.22	2.19	2.15	2.12	2.09	2.07
	.01	2.11	2.07	2.04	2.01	1.98	1.95
500	.05	1.69	1.66	1.64	1.62	1.61	1.59
	.01	2.07	2.04	2.00	1.97	1.94	1.92
1000	.05	1.68	1.65	1.63	1.61	1.60	1.58
	.01	2.06	2.02	1.98	1.95	1.92	1.90

Source: Table prepared by the author using Microsoft Excel's *F* inverse function. (Excel is a registered trademark of the Microsoft Corporation.)

Table 5 Values of the Studentized Range Statistic, q_k

Note: For a one-way ANOVA the value of k is the number of means in the factor.

Critical values for $\alpha = .05$ in regular text
Critical values for $\alpha = .01$ in **bold text**

Degrees of Freedom within Groups $(df_{wn} = N - k)$	α	k = Number of Means Being Compared					
		2	**3**	**4**	**5**	**6**	**7**
1	.05	18.00	27.00	32.80	37.10	40.40	43.10
	.01	**90.00**	**135.0**	**164.0**	**186.0**	**202.0**	**216.0**
2	.05	6.09	8.30	9.80	10.90	11.70	12.40
	.01	**14.00**	**19.00**	**22.30**	**24.70**	**26.60**	**28.20**
3	.05	4.50	5.91	6.82	7.50	8.04	8.48
	.01	**8.26**	**10.60**	**12.20**	**13.30**	**14.20**	**15.00**
4	.05	3.93	5.04	5.76	6.29	6.71	7.05
	.01	**6.51**	**8.12**	**9.17**	**9.96**	**10.60**	**11.10**
5	.05	3.64	4.60	5.22	5.67	6.03	6.33
	.01	**5.70**	**6.97**	**7.80**	**8.42**	**8.91**	**9.32**
6	.05	3.46	4.34	4.90	5.31	5.63	5.89
	.01	**5.24**	**6.33**	**7.03**	**7.56**	**7.97**	**8.32**
7	.05	3.34	4.16	4.69	5.06	5.36	5.61
	.01	**4.95**	**5.92**	**6.54**	**7.01**	**7.37**	**7.68**
8	.05	3.26	4.04	4.53	4.89	5.17	5.40
	.01	**4.74**	**5.63**	**6.20**	**6.63**	**6.96**	**7.24**
9	.05	3.20	3.95	4.42	4.76	5.02	5.24
	.01	**4.60**	**5.43**	**5.96**	**6.35**	**6.66**	**6.91**
10	.05	3.15	3.88	4.33	4.65	4.91	5.12
	.01	**4.48**	**5.27**	**5.77**	**6.14**	**6.43**	**6.67**
11	.05	3.11	3.82	4.26	4.57	4.82	5.03
	.01	**4.39**	**5.14**	**5.62**	**5.97**	**6.25**	**6.48**
12	.05	3.08	3.77	4.20	4.51	4.75	4.95
	.01	**4.32**	**5.04**	**5.50**	**5.84**	**6.10**	**6.32**
13	.05	3.06	3.73	4.15	4.45	4.69	4.88
	.01	**4.26**	**4.96**	**5.40**	**5.73**	**5.98**	**6.19**
14	.05	3.03	3.70	4.11	4.41	4.64	4.83
	.01	**4.21**	**4.89**	**5.32**	**5.63**	**5.88**	**6.08**
16	.05	3.00	3.65	4.05	4.33	4.56	4.74
	.01	**4.13**	**4.78**	**5.19**	**5.49**	**5.72**	**5.92**

Degrees of Freedom within Groups $(df_{wn} = N - k)$	α	k = Number of Means Being Compared					
		2	**3**	**4**	**5**	**6**	**7**
18	.05	2.97	3.61	4.00	4.28	4.49	4.67
	.01	**4.07**	**4.70**	**5.09**	**5.38**	**5.60**	**5.79**
20	.05	2.95	3.58	3.96	4.23	4.45	4.62
	.01	**4.02**	**4.64**	**5.02**	**5.29**	**5.51**	**5.69**
24	.05	2.92	3.53	3.90	4.17	4.37	4.54
	.01	**3.96**	**4.54**	**4.91**	**5.17**	**5.37**	**5.54**
30	.05	2.89	3.49	3.84	4.10	4.30	4.46
	.01	**3.89**	**4.45**	**4.80**	**5.05**	**5.24**	**5.40**
40	.05	2.86	3.44	3.79	4.04	4.23	4.39
	.01	**3.82**	**4.37**	**4.70**	**4.93**	**5.11**	**5.27**
60	.05	2.83	3.40	3.74	3.98	4.16	4.31
	.01	**3.76**	**4.28**	**4.60**	**4.82**	**4.99**	**5.13**
120	.05	2.80	3.36	3.69	3.92	4.10	4.24
	.01	**3.70**	**4.20**	**4.50**	**4.71**	**4.87**	**5.01**
	.05	2.77	3.31	3.63	3.86	4.03	4.17
	.01	**3.64**	**4.12**	**4.40**	**4.60**	**4.76**	**4.88**

Note: For a one-way ANOVA the value of *k* is the number of means in the factor.

Critical values for α = .05 in regular text
Critical values for α = .01 in **bold text**

Degrees of Freedom within Groups ($df_{wn} = N - k$)	α	8	9	10	11	12
				k = Number of Means Being Compared		
1	.05	45.40	47.40	49.10	50.60	52.00
	.01	**227.0**	**237.0**	**246.0**	**253.0**	**260.0**
2	.05	13.00	13.50	14.00	14.40	14.70
	.01	**29.50**	**30.70**	**31.70**	**32.60**	**33.40**
3	.05	8.85	9.18	9.46	9.72	9.95
	.01	**15.60**	**16.20**	**16.70**	**17.10**	**17.50**
4	.05	7.35	7.60	7.83	8.03	8.21
	.01	**11.50**	**11.90**	**12.30**	**12.60**	**12.80**
5	.05	6.58	6.80	6.99	7.17	7.32
	.01	**9.67**	**9.97**	**10.20**	**10.50**	**10.70**
6	.05	6.12	6.32	6.49	6.65	6.79
	.01	**8.61**	**8.87**	**9.10**	**9.30**	**9.49**
7	.05	5.82	6.00	6.16	6.30	6.43
	.01	**7.94**	**8.17**	**8.37**	**8.55**	**8.71**
8	.05	5.60	5.77	5.92	6.05	6.18
	.01	**7.47**	**7.68**	**7.87**	**8.03**	**8.18**
9	.05	5.43	5.60	5.74	5.87	5.98
	.01	**7.13**	**7.32**	**7.49**	**7.65**	**7.78**
10	.05	5.30	5.46	5.60	5.72	5.83
	.01	**6.87**	**7.05**	**7.21**	**7.36**	**7.48**
11	.05	5.20	5.35	5.49	5.61	5.71
	.01	**6.67**	**6.84**	**6.99**	**7.13**	**7.26**
12	.05	5.12	5.27	5.40	5.51	5.62
	.01	**6.51**	**6.67**	**6.81**	**6.94**	**7.06**
13	.05	5.05	5.19	5.32	5.43	5.53
	.01	**6.37**	**6.53**	**6.67**	**6.79**	**6.90**
14	.05	4.99	5.13	5.25	5.36	5.46
	.01	**6.26**	**6.41**	**6.54**	**6.66**	**6.77**
16	.05	4.90	5.03	5.15	5.26	5.35
	.01	**6.08**	**6.22**	**6.35**	**6.46**	**6.56**

Degrees of Freedom within Groups $(df_{wn} = N - k)$	α	k = Number of Means Being Compared				
		8	9	10	11	12
18	.05	4.82	4.96	5.07	5.17	5.27
	.01	**5.94**	**6.08**	**6.20**	**6.31**	**6.41**
20	.05	4.77	4.90	5.01	5.11	5.20
	.01	**5.84**	**5.97**	**6.09**	**6.19**	**6.29**
24	.05	4.68	4.81	4.92	5.01	5.10
	.01	**5.69**	**5.81**	**5.92**	**6.02**	**6.11**
30	.05	4.60	4.72	4.83	4.92	5.00
	.01	**5.54**	**5.56**	**5.76**	**5.85**	**5.93**
40	.05	4.52	4.63	4.74	4.82	4.91
	.01	**5.39**	**5.50**	**5.60**	**5.69**	**5.77**
60	.05	4.44	4.55	4.65	4.73	4.81
	.01	**5.25**	**5.36**	**5.45**	**5.53**	**5.60**
120	.05	4.36	4.48	4.56	4.64	4.72
	.01	**5.12**	**5.21**	**5.30**	**5.38**	**5.44**
	.05	4.29	4.39	4.47	4.55	4.62
	.01	**4.99**	**5.08**	**5.16**	**5.23**	**5.29**

Source: Abridged from H. L. Harter, D. S. Clemm, and E. H. Guthrie, The probability integrals of the range and the studentized range, WADC Technical Report 58-484, Vol. 2, 1959, Wright Air Development Center, Table II. 2, pp. 243–281. Reprinted from B.J. Winer, *Statistical Principles in Experimental Design*, 2E (1965), The McGraw-Hill Companies. Reproduced with permission of the McGraw-Hill Companies.

GLOSSARY OF STATISTICAL FORMULAS

CHI SQUARE

$$X^2 = \sum \left(\frac{(f_o - f_e)^2}{f_e} \right)$$

CONFIDENCE INTERVAL

$$(S_{\bar{X}})(-t_{\text{crit}}) + \overline{X} < \mu < (S_{\bar{X}})(+t_{\text{crit}}) + \overline{X}$$

CORRELATION

Pearson's Product Moment Correlation Coefficient

$$r = \frac{N\left(\sum XY\right) - \left(\sum X\right)\left(\sum Y\right)}{\sqrt{\left[N\left(\sum X^2\right) - \left(\sum X\right)^2\right]\left[N\left(\sum Y^2\right) - \left(\sum Y\right)^2\right]}}$$

Spearman's rho

$$r_s = 1 - \frac{6\left(\sum D^2\right)}{N(N^2 - 1)}$$

DEVIATIONS FROM THE MEAN

$$D = X - \bar{X}$$

EFFECT SIZE
Phi Coefficient (Chi-Square)

$$\Phi = \sqrt{\frac{X_{obt}^2}{N}}$$

Point Biserial Correlation Coefficient (Independent and Related Samples *t* Test)

$$r_{pb}^2 = \frac{(t_{obt})^2}{(t_{obt})^2 + df}$$

Eta Squared (One Way ANOVA)

$$\acute{\eta}^2 = \frac{SS_{bn}}{SS_{tot}}$$

MEAN

$$\bar{X} = \frac{\sum X}{n}$$

MEDIAN LOCATOR

$$\frac{n+1}{2}$$

ONE WAY ANALYSIS OF VARIANCE

Sum of Squares Total $= SS_{\text{tot}} = \sum X_{tot}^2 - \dfrac{\left(\sum X_{tot}\right)^2}{N}$

Sum of Squares Within Groups $= SS_{\text{wn}} = \sum \left[\sum X^2 - \dfrac{\left(\sum X\right)^2}{n} \right]$

Sum of Squares Between Groups $= SS_{\text{bn}} = \sum \left[\dfrac{\left(\sum X\right)^2}{n} \right] - \dfrac{\left(\sum X_{tot}\right)^2}{N}$

Degrees of Freedom Total $= df_{\text{tot}} = N{-}1$
Degrees of Freedom Within Groups $= df_{\text{wn}} = N{-}k$
Degrees of Freedom Between Groups $= df_{\text{bn}} = k{-}1$

Mean Squares Between Groups $= MS_{\text{bn}} = \dfrac{SS_{bn}}{df_{bn}}$

Mean Squares Within Groups $= MS_{\text{wn}} = \dfrac{SS_{wn}}{df_{wn}}$

F Statistic $= F = \dfrac{MS_{bn}}{MS_{wn}}$

Tukey HSD $= (q_{\text{k}})\left(\sqrt{\dfrac{MS_{wn}}{n}} \right)$

PERCENTAGE

$$\% = \frac{f}{N} \times 100$$

POOLED VARIANCE

$$V_{\text{pool}} = \frac{(n_1 - 1)(v_1) + (n_2 - 1)(v_2)}{(n_1 - 1)(n_2 - 1)}$$

PROPORTION

$$p = \frac{f}{N}$$

RANGE

$$H-L$$

REGRESSION

| Slope *(b)* | *y* intercept *(a)* | Regression Equation |

$$b = \frac{N\left(\sum XY\right) - \left(\sum X\right)\left(\sum Y\right)}{N\left(\sum X^2\right) - \left(\sum X\right)^2}$$

$$a = \bar{Y} - (b)\bar{X}$$

$$Y' = b(X) + a$$

STANDARD DEVIATION

Computational Formulas

$$S = \sqrt{\frac{\sum X^2 - \frac{\left(\sum X\right)^2}{N}}{N}} \text{ (for sample)}$$

$$\sigma = \sqrt{\frac{\sum X^2 - \frac{\left(\sum X\right)^2}{N}}{N}} \text{ (for population)}$$

$$s = \sqrt{\frac{\sum X^2 - \frac{\left(\sum X\right)^2}{N}}{N - 1}} \text{ (estimated population standard deviation)}$$

Definitional Formulas

$$S = \sqrt{\frac{\sum (X - \bar{X})^2}{N}} \text{ (for sample)}$$

$$\sigma = \sqrt{\frac{\sum (X - \bar{X})^2}{N}} \text{ (for population)}$$

$$s = \sqrt{\frac{\sum (X - \bar{X})^2}{N - 1}} \text{ (estimated population standard deviation)}$$

STANDARD ERROR
Standard Error of the Estimate

$$Sy' = Sy\sqrt{1 - r^2}$$

Standard Error of the Mean

$$\delta_{\bar{X}} = \frac{\sigma}{\sqrt{n}} \text{ (for } z \text{ test)} \qquad s_{\bar{X}} = \sqrt{\frac{v}{N}} \text{ (for } t \text{ test) (estimated standard error of the mean)}$$

Standard Error of the Mean Difference

$$v_{\bar{X}_1 - \bar{X}_2} = \sqrt{(V_{pool})\left(\frac{1}{n_1} + \frac{1}{n_2}\right)} \text{ (independent samples } t \text{ test)}$$

Standard Error of the Mean Difference

$$v_{\bar{D}} = \sqrt{\frac{v_D}{N}} \text{ (related samples } t \text{ test)}$$

t STATISTIC

$$t = \frac{\bar{X} - \mu}{S_{\bar{X}}} \text{ (one sample } t \text{ test)}$$

$$t = \frac{\bar{X}_1 - \bar{X}_2}{v_{\bar{X}_1 - \bar{X}_2}} \text{ (independent samples } t \text{ test)}$$

$$t = \frac{\bar{D}}{v_{\bar{D}}} \text{ (related samples } t \text{ test)}$$

VARIANCE

$$V = \frac{\sum X^2 - \frac{\left(\sum X\right)^2}{N}}{N} \quad \text{(for sample)} \qquad \sigma^2 = \frac{\sum X^2 - \frac{\left(\sum X\right)^2}{N}}{N} \quad \text{(for population)}$$

$$v = \frac{\sum X^2 - \frac{\left(\sum X\right)^2}{N}}{N - 1} \quad \text{(estimated population variance)}$$

$$v_D = v^2 = \frac{\sum D^2 - \frac{\left(\sum D\right)^2}{N}}{N} \quad \text{(estimated variance of the difference)}$$

Z SCORES

$$z = \frac{X - \bar{X}}{S} \quad \text{(for raw scores)} \qquad z = \frac{\bar{X} - \mu}{\sigma_{\bar{X}}} \quad \text{(for sample means)}$$

Determining Raw Scores from *z* Scores

$$X = (z)(S) + \bar{X}$$

INDEX

Note: Page numbers followed by *f* and *t* indicate figures and tables